ON THE EDGE: MAPPING NORTH AMERICA'S COASTS

On the Edge: Mapping
North America's Coasts

Roger M. McCoy

OXFORD
UNIVERSITY PRESS

OXFORD
UNIVERSITY PRESS

Oxford University Press, Inc., publishes works that further
Oxford University's objective of excellence
in research, scholarship, and education.

Oxford   New York
Auckland   Cape Town   Dar es Salaam   Hong Kong   Karachi
Kuala Lumpur   Madrid   Melbourne   Mexico City   Nairobi
New Delhi   Shanghai   Taipei   Toronto

With offices in
Argentina   Austria   Brazil   Chile   Czech Republic   France   Greece
Guatemala   Hungary   Italy   Japan   Poland   Portugal   Singapore
South Korea   Switzerland   Thailand   Turkey   Ukraine   Vietnam

Copyright © 2012 by Oxford University Press

Published by Oxford University Press, Inc.
198 Madison Avenue, New York, New York 10016

www.oup.com

Library of Congress Cataloging-in-Publication Data
McCoy, Roger M.
On the edge : mapping North America's coasts / Roger M. McCoy.
p.   cm.
Includes bibliographical references and index.
ISBN 978-0-19-974404-6 (hardcover : alk. paper) 1. Cartography—Atlantic Coast
(North America)—Early works to 1800. 2. Cartography—Pacific Coast (North America)—
Early works to 1800. 3. Atlantic Coast (North America)—Maps—Early works to 1800.
4. Pacific Coast (North America)—Maps—Early works to 1800. I. Title.
GA401.M43   2012
526.0970914'6—dc23   2012005798

1 3 5 7 9 8 6 4 2

Printed in the United States of America
on acid-free paper

To the incredible courage and unwavering determination that impels adventurers through the perils of unknown lands in pursuit of a dream. To adventurers Sue, Michael, Sherry, Claire, Max, and Leah.

We were wanderers on a prehistoric earth, on an earth that wore the aspect of an unknown planet. We could have fancied ourselves the first of men taking possession of an accursed inheritance, to be subdued at the cost of profound anguish and of excessive toil. . . . The dreams of men, the seed of commonwealths, the germs of empires.

—JOSEPH CONRAD

Table of Contents

Acknowledgments

A special mention of gratitude is owed to Professor Dennis Fitzsimons, Humboldt State University, for his generous advice and effort toward improving the maps in this book. His expertise has been invaluable. I also wish to acknowledge the help given by Nancy Lemay, map librarian at the University of Ottawa, and Kristen Delorey of Natural Resources Canada for their helpful information on Canadian maps and aerial photos. Others that provided valuable help were Joan Fitzpatrick of the U. S. Geological Survey, Tom Weingarten of the University of Alaska Institute of Marine Science, and Humfrey Melling of the Canada Institute of Ocean Science. Sue McCoy gave a huge piece of her time to reading and correcting drafts. Her insightful questions, comments, and suggestions were always constructive and greatly appreciated.

ON THE EDGE: MAPPING NORTH AMERICA'S COASTS

PART I

The Earliest Ventures to North America to Find a Northwest Passage, 1492–1543

I am rather inclined to believe that this is the land God gave to Cain.
—JACQUES CARTIER (first impression of the north shore of St. Lawrence Gulf)

1

The Urge to Discover New Lands and Make Maps

⌐ ——

SUPPOSE THE SEARCH for the Northwest Passage had begun around AD 1000—
at the time Norsemen were pushing into the North American Arctic waters. Norse
mariners had the good fortune to sail in a warm period during which the northern
seas may have been less obstructed by ice. An interval of almost five hundred years
elapsed between the Norse settlement on Newfoundland and John Cabot's redis-
covery of it in 1497 while seeking a passage to Asia. During those five hundred years,
important climatic changes occurred. Cabot's voyage occurred in the early years of a
profound climatic cooling heralding the beginning of the Little Ice Age. Recent data
from ice cores in Greenland substantiate a great cooling from the mid-fifteenth
century to the mid-nineteenth century.[1]

The long period of colder climate coincided with the avid interest of English and
French monarchs and merchants to find a shorter, northern route to Asia and avoid
the long route around south Africa that was controlled by the Portuguese and
Spanish. By the time English explorers began sailing to North America in the sixteenth century, the Little Ice Age had a firm grip on polar regions. All exploration
for the Northwest Passage, up through the middle of the nineteenth century, took
place during this period of icy climate with very short sailing seasons when ice floes
blocked much travel. Some sea routes could be open one year and choked by ice the
next. One year an expedition could fail and be followed a year later by another expedition sailing easily through the same location. However, none sailed through the all
important passage to the Pacific Ocean.

The earliest European mariners entering the Arctic seas had little or no experience sailing in extremely cold climates. They faced icebergs towering over their small ships (the caravel's main deck was about 60 feet by 18 feet). If a sudden shifting of wind caused the ice pack to move, shutting off a route before and behind them, mariners could be suddenly surrounded by ice. Strong tides could push them into an ice mass, their wooden hulls crushed easily by the pressure of converging ice floes squeezing the ship. Navigators could not rely on their compasses because the magnetic field behavior seemed more erratic in these waters than in the home waters of Europe's west coast. Even in the summer sailing season, storms would coat the sails and rigging with ice. Expeditions often spent the winter in protected bays where they became locked in ice but avoided the crushing pressures of the immense ice floes in open water, where the ice was constantly moving. After wintering in an ice-locked bay, sailors often had to cut channels through the ice with saws and axes and tow their ships to open water where the summer's melting created an ice-free sea.

The Europeans' fertile imaginations provided ample incentive to push farther into the unknown. Rumors persisted about waterways connecting the Atlantic Ocean to the Pacific. So certain were the fifteenth- and sixteenth-century cartographers that such a route must exist, some drew an imagined passage on their maps calling it the Strait of Anian, probably named for the Chinese province of Ania, mentioned by Marco Polo. Two mariners, Juan de Fuca (1592) and Bartholomew de Fonte (1640), even claimed to have sailed the Strait. In 1562, the Strait of Anian first appeared on a map drawn by Italian cartographer Giacomo Gastaldi. The fictional strait grew in the European imagination, convincing sailors to search for the sea route from Europe to China. Some maps showed the Strait of Anian extending from approximately Puget Sound to Hudson Bay, others from the Sea of Cortez to Hudson Bay. The Spanish, Portuguese, French, and English probed every waterway, large river, and bay, expecting each to be the elusive passage to the Pacific. Every expedition in the sixteenth century began with the expectation of finding it.

The English, directing their attention to the North Atlantic, began to map all the coasts they encountered. Many places in the Arctic region are known by names of English explorers, such as Davis Strait, Baffin Island, Frobisher Bay, Hudson Bay, Foxe Basin, and James Bay. English names label almost every waterway and island in the Canadian Archipelago, recalling the names of the mariners, their ships, or their benefactors.

Hopes for a Northwest Passage led many promoters to jump at any small shred of information as the basis for proposing yet another voyage of exploration. They scoured the logs of previous voyages for information on tides and currents that might indicate a possible opening to an unseen ocean. They seized on accounts of mariners who boasted of finding the opening to a passage that had not been explored. Promoters were always hoping to convince investors and monarchs that the next voyage would lead to discovery of the elusive passage.

The value of these early voyages should not be measured by their successes, for none of the explorers found what they sought. But each added important information to the understanding of the North American continent they encountered. Their maps expanded the extent of knowledge about the size and shape of the continent, and each expedition provided impetus for others to come in search of the passage. Each journey contributed to the realization that an open, navigable, northern passage to China did not exist in either the Western or in the Eastern Hemispheres. The fact that previous explorers had not found a northern passage was assumed to mean only that they had not looked in the right places. So the effort had to continue until every inlet, bay, and strait had been tested. Their efforts spanned four hundred years and hundreds of searchers before one small ship, commanded by Roald Amundsen, sailed from the Atlantic to the Pacific across the top of North America (taking three seasons from 1903–1906).

THE EARLY SURGE IN MAPPING AND EXPLORATION

When the fifteenth-century Europeans arrived in North America, they brought with them their own perception of the world. They imagined they had arrived in the "Indies," a vague name that included a vast area of Asia. The nebulous area they imagined stretched, in their world perception, from China to Turkey and included India and Indonesia. By the early part of the sixteenth century, the Europeans began to realize they were not yet in the Indies of their imagination. The land masses of two enormous continents stood in the way and were first regarded as a barrier to their ambitions. The explorers immediately began to look for a route around these lands so they could proceed on to Asia.

Only twenty-eight years after Columbus's first voyage in 1492, a Spanish expedition commanded by the Portuguese mariner, Magellan, sailed around the south end of South America and found another ocean west of the intruding land mass. Within fifty years, from 1492 to 1542, explorers had seen and mapped both coasts of South America, the coasts of North America to Labrador on the east, and the west coast to Cape Blanco, Oregon. The North American portion known after this rapid rate of exploration is portrayed in the map of 1543 (figure 1.1).

Following this astonishing fifty-year period of exploring and mapping American coastlines, the process slowed considerably. Completion of the map of the northernmost parts of North America required almost 350 years of repeated effort with many failures in the attempt. Geographic logic of that time demanded that a passage in the south would be matched by one in the north.

The discovery of continents in the Western Hemisphere led to such intensive exploration that the author George Best, who accompanied Martin Frobisher and wrote an account of Frobisher's late sixteenth-century voyages, said that more regions and countries had been discovered in the previous eighty years than in the past five thousand years. Furthermore, he wrote in 1578 that more than half the world had been discovered by men then living.[2] We hear similar statements in our time regarding the great number of recent scientific discoveries.

George Best continued by saying that the skills for "the use of the compass, the houre-glasse for observing time, instrumentes of astronomie to take longitudes and latitudes of countreys, and many other helps, are so commonly knowen of every

FIGURE 1.1 The heavy line represents the known coastlines of North America up to 1543.

mariner now adayes, that he that hath bin twice at sea, is ashamed to come home if he be not able to render accounte of all these particularities." George Best may not have realized that the complex astronomical observations and computations required to determine longitude often resulted in great locational errors.

Portugal and Spain made the first claims on land in the new world continents, and with the Treaty of Tordesillas in 1494 divided the remainder of the Western Hemisphere as their domain. This ambitious treaty, which was sanctioned by Pope Alexander VI and modified in 1506 by Pope Julius II, presented little hindrance to English merchants, who also had strong interest in sea routes to Asia. England could not directly challenge the dominance of the Spanish and Portuguese powers and their established southern routes to the Orient around Africa. However, the English crown, not being a party to the treaty, felt no constraint in searching for a northerly route, even though the terms of the treaty gave them no rights in that vast area. Despite the pope's decree, the Englishman George Best said in 1578, "Yea, there are countries yet remaining without masters and possessors . . . in which lands lack nothing that may be desired either for pleasure, profit, or necessary uses. To possess and obtain sundry countries is an easy thing, so I would not have our English nation to be slack therein." Best made clear his feeling that England had been "slack" in letting Spain dominate the West Indies.

European explorers in the fifteenth century considered all lands not already ruled by a Christian prince to be open to claim for their own country. Claiming a newly discovered land required a formal ceremony on shore. The ship's captain represented his king and read a declaration defining the extent of land and the name of its new ruler. Usually a monument or cross was erected. If any natives of the land were present, they were warned in a language they could not understand not to commit any treasonous acts against the king lest they be severely punished.

Maps based on information provided by navigators were essential for claiming the land. The country claiming new land needed to show where their claim was located and its extent if they expected other countries to honor their claim. If, as sometimes occurred, a country chose to keep its maps secret, its claimed land would often be taken by another claimant. Claims were often ignored when a vast area, such as an entire continent or ocean, was claimed. In 1513, Balboa waded into the Pacific Ocean claiming the land and all shores around it, and for centuries after that, the Spanish considered the Pacific with all islands and adjacent lands to be theirs. Other countries paid little attention to such a grandiose claim.

The English hoped to find a route that avoided the Portuguese and Spanish sea routes completely. Although a few efforts were made to find a northeast route north of Russia, that prospect soon led to disappointment when the explorers found the only possible route was perpetually blocked by ice.

The northern parts of North America, on the other hand, had so many alternative sea passages among the islands, inlets, and bays that scores of voyages over 350 years were made before all hope was quashed. As the English and French mariners brought back news of their discoveries and failures, maps of North America evolved to show the form of the continent—sometimes including that fictitious seaway connecting the Atlantic and Pacific Oceans. Many expeditions were inspired solely by the imagination of some mapmaker whose mind envisioned the Northwest Passage, convinced that someone would soon discover it.

The Florentine explorer, Giovanni da Verrazzano, sailed the North American coast from France in 1524 looking for a passage. Sailing past a part of the North Carolina coast, he assumed the water of Pamlico Sound beyond the barrier bars was an ocean connecting to Asia. This error, known as the Sea of Verrazzano, appeared on maps for a hundred years.

Another optimistic and imaginative example was Mercator's map of the world in 1538 showing open seas up to the North Pole, based on the belief that sea ice would be found only near land where water was less salty. In 1576, Humphrey Gilbert showed the St. Lawrence River extending all the way to the Gulf of California. Martin Frobisher, the explorer bitten by gold fever in 1576, declared that what is now called Frobisher Bay, on Baffin Island, was actually a strait extending across Canada to a western ocean. In 1585, John Davis thought the strait now named for him was the beginning of the Northwest Passage. If he had continued a bit farther he would have discovered the opening that was indeed the actual beginning of the Passage!

DEVELOPMENT OF MAPS OF NORTH AMERICA

Maps made in the fifteenth and sixteenth centuries appear distorted, often beyond recognition to our eyes today. This distortion can be attributed to two main causes. First, the mariners could not measure longitude accurately. They simply had to estimate their position east or west from some starting point of known longitude. Their best effort at estimating longitude used observations of time, speed, and direction to estimate the distance traveled on their intended course. Observations of both speed and time were estimates and were subject to errors, and the farther they traveled, the more the errors might accumulate. Therefore, when a mariner recorded the location of a coastline, he was often wrong by hundreds of miles in an east–west direction.

Latitude, on the other hand, could be measured fairly accurately by the fifteenth century by measuring the angle to the noon sun or to a "guard" star of the Little Dipper, and later to the north star, Polaris. Thus we usually find old maps fairly reliable in a north–south direction but distorted east and west. One well-known example is Frobisher Bay in Baffin Island in the Canadian Arctic, which a sixteenth-century map shows as being 700 miles to the east in southern Greenland—but at the correct latitude.

A second reason for distortion is the imagination of the mapmaker. If the cartographer wanted to promote the idea that a sea passage existed from Europe to Asia, he would simply design his map to show its presence. Where a lack of information existed, the mapmaker could fill in those gaps in any way. Hence, some very old maps show a passage to Asia, and some show North America connected to Asia. The cartographers' imaginary creations often had names of mythological places, such as Gog and Magog, a place thought to be inhabited by giants and supernatural beings, or Quivera or Cibola, two of the seven mythical cities of gold. The thinking of fifteenth-century map makers about accuracy was vastly different from today's cartographers who believe "If you don't know, don't show."

The exploring mariners recorded their routes and made notes and sketches of what they saw, but often had little to say about the final map produced by a cartographer. The voyage of discovery was made under the sanction of the crown, and typically the notes and logs from the trip were turned over to a designated cartographer who used the information to create a map. In a few instances, a mariner with adequate skills created his own map.

The map became property of the king. Some countries carefully guarded the maps showing new territories and routes they had discovered, just as a business kept trade secrets. Portugal particularly did not want others to know details of their route to Asia via Africa's Cape of Good Hope. Some countries made maps available only to selected trading partners.

The cartographer who most ignited the fifteenth-century mariners' imagination had lived thirteen centuries earlier. Claudius Ptolomy (ca. 85–ca. AD 168) lived in Alexandria, Egypt, where he used the famous library to compile the known knowledge of astronomy, geography, and astrology into three books. The astronomy and geography books had a long-lasting influence, but they both presented serious errors that went uncorrected for about 1,300 years. Ptolemy's astronomy book, *Almagest*, rejected the theory earlier proposed by Aristarchus (ca. 230 BC) that the earth revolves around the sun. Ptolemy's geocentric idea accepted the ideas of Aristotle and formed the main thesis of his book. When Ptolemy's works were rediscovered, after being lost throughout the Middle Ages, they were accepted as gems of ancient wisdom, and few had the nerve or the authority to challenge them. Galileo, for example, paid a price for presenting his heliocentric ideas. Likewise, any sixteenth-century maps that altered the venerable Ptolemy map were regarded with suspicion.

In his other book, *Geographia*, Ptolemy rejected Eratosthenes' (d. 195 BC) nearly correct computation of the circumference of the earth. Rather, he chose an erroneous and much smaller distance (about 75 percent of the actual size). Ptolemy did not make any measurements himself, as Eratosthenes had done, but selectively compiled other information that was known at the time. The estimate he chose came from the Greek astronomer Poseidonius. Subsequently, however, his choices became known as Ptolemaic ideas and were considered irrefutable. Also, Ptolemy assumed that the known world's land surface covered 180° of longitude ranging from the Canary Islands in the west to the easternmost part of Asia (about 20° of longitude too much). This error on his map showed the Atlantic Ocean much too narrow between western Europe and east Asia, without the American continents in between. This remained the understanding of the world for 1,300 years.

The combination of these two errors—Earth's circumference too small and land area too large—encouraged mariners of the fifteenth century to assume that a relatively short voyage across the Atlantic Ocean would take them to Asia. Columbus was the first to promote an expedition on the basis of these errors. When Columbus made landfall, he had traveled as far as he expected to travel to reach Asia and logically assumed that he had succeeded.

To his credit, Ptolemy's map introduced some excellent standards to cartography, despite the errors. Though he was not the first to use the idea of a gridded coordinate system, his method of showing latitude and longitude became a standard for future maps. Also, Ptolomy insisted that maps should be drawn to scale. Many maps of his time were distorted by enlarging the better known places in order to include all the known information. Unfortunately, many cartographers of his time failed to adopt his practical approach to scale and location. Most simply fitted their maps to the size of some convenient boundary line drawn on a piece of parchment.

Maps began to proliferate in the sixteenth century. Each voyage of exploration and discovery provided new information that had to be mapped. In 1507, a German cartographer, Martin Waldseemüller, produced a map of the world, *Universalis Cosmographia*, which was the first to show Columbus's discovery as a separate continent. But he cautiously made the new continent very narrow—just a long, skinny island—rather than contradict Ptolemy's erroneous circumference of Earth. By the middle of the sixteenth century, enough voyages had been made, including Magellan's trip around the world, that most cartographers recognized that Earth's circumference shown on Ptolemy's map was wrong.

In 1569, the Flemish cartographer, Gerardus Mercator, made a map of the world showing all the known lands using his now famous innovative grid system of latitude and longitude. In 1570, Abraham Ortelius, a Belgian cartographer, made the first known atlas of the world in an effort to compile the rapidly accumulating geographic knowledge. Although these maps still had some residual traits of the Ptolemy map, they showed great improvements in detail and accuracy. A major feature retained from the Ptolemy map was the presence of a very large continent in the Antarctic—large enough to counterbalance the immense weight of the land areas of the Northern Hemisphere. This belief, dating back to Aristotle, flowed from the Greek concept of symmetry, but was also based on the thought that Earth would not turn smoothly if not in perfectly balanced equilibrium. This view of an immense Antarctica persisted until 1768 when Captain James Cook's voyage in the *Endeavour* showed that the Antarctic continent indeed existed, but was much smaller than previously believed.

In the end, the dreamers of a Northwest Passage were right. Just as they had found a sea passage around the southern continent, there was also a passage in the north. They were right to say that land does not extend all the way to the North Pole. What they had not imagined was that the passage was so far north and that it would be obstructed by a mass of ice completely impenetrable for their light sailing vessels. Only after about four hundred years of effort, along with important technological improvements in ships, could anyone at last navigate the Northwest Passage.

Despite many failures, early mariners returned to Europe with detailed records of their voyages and with their best estimates of latitude and longitude for every land feature they had seen. Maps of Arctic lands improved with each voyage, and gradually the coastline of North America and all its Arctic islands began to appear on maps.

Only by keeping a record of their navigation results could mariners create a map of the new lands they had encountered. As they sailed along a coast, they recorded latitude and estimated longitude at every change in the shoreline. For greatest accuracy, they would anchor their ship and go ashore to stand on a stable surface to get the best results for the observation of stellar and solar angles. If the navigator had some cartographic skills, he could plot the points on a chart, sketching the configuration of the coast to include embayments, promontories, or river mouths that occurred between the recorded locations. The magnitude of a map's longitudinal error depended on the navigators' success in estimating their westward progress by means of dead reckoning. The use of the lunar distance method of computing longitude was known by the early sixteenth century, but few mariners had the mathematical skill to use it. The lunar distance method was likewise prone to error from several sources. Maps made by the early cartographers are so full of erroneous information and outright falsehoods that it is difficult to understand just how much the explorers actually achieved. Many readers today have difficulty knowing what was real and what was imaginary on those old maps.

Maps made for this book somewhat clarify this problem by showing all information at the same scale and projection. A sequence of maps shows the stages of the world's awareness of North American coasts, depicting what mariners actually achieved, rather than how their maps looked with various distortions, errors, and imaginary creations.

In addition to the maps, accounts of selected voyages are provided, telling of the successes and mishaps during those four hundred years. These accounts are based, whenever possible, on the narratives of the explorers themselves. The act of traveling in a small sailing ship to unknown lands, especially the Arctic, required great courage, even if everything worked as planned. However, in many voyages, almost nothing went as planned, and ships and lives were lost. These stories of the mariners are necessarily a part of appreciating the value of the maps as they show the emerging understanding of the North American coast lines. Their stories show the dangers and hardships men endured while completing the maps. A chronology for all expeditions mentioned in this book is provided in the appendix.

The discovery of most of North America was motivated and initiated by the commercial interests of merchants in England and France as well as the imperial ambitions of those two countries. The immense effort to expand commerce and imperial power required, first of all, charting the coasts and waterways of all new territories. Maps, beginning with the earliest incomplete charts of John Cabot in 1497, were filled in and eventually completed by all the subsequent explorers, each of whom added detail in known areas and expanded into new lands. Dozens of expeditions over the years gradually filled in the details of an unknown continental landmass. The chart makers provided more than coastal outline; they also included information related to sailing conditions. Every bit of information added to the maps on harbors, shoals, hazardous rocks, and ice conditions provided greater certainty for explorers who came later.

In addition, the narrative written by explorers was as important as the chart itself. Without the narrative, a voyage might soon be forgotten, and a map would be the only vestige of an expedition. The narrative established the new land as a real place—not an abstraction. It provided details of the flora, fauna, resource potential, and the human inhabitants, along with the trials and successes of the voyage. Maps, with their narratives, transformed a land from an unknown place to a known entity—a possession with a known owner—for the world to see.

Many expedition narratives are beautifully written and illustrated and are a pleasure to read, and they often became widely read among the contemporary public, eager to hear about the strange new lands. The illustrations were not merely for decoration but were an authentication of the narrative, as an eyewitness account. The illustrations attached a view of reality to the map. They complement the purpose of the narrative by emphasizing that the map is not merely a field of coordinates, but that each coordinate stands for a place. The illustrations held such importance that a lack of artistic ability came to be considered a serious handicap for an explorer.

Some member of an expedition had to be skilled with pencil and water colors.[3] This was particularly true for the nineteenth-century voyages of exploration by the English navy.

When explorers began to sail to the far north, their maps and the narratives that accompanied them gave the public a view of the Arctic as a place of desolation with dangerous ice floes fraught with occasional disasters. People were fascinated by the adventures of these courageous men in a region that had little potential for settlement or commerce. The lack of any apparent utility in the Arctic was offset during the nineteenth century by the public's fascination with exploration itself, mixed with pride of empire for an England that remained at the forefront of important events. It is not hard to understand the public view when we recall a similar fascination with men landing on the moon in the twentieth century and the resulting national pride felt by most Americans.

The nineteenth-century maps of North America were usually produced by representatives of the government, members of the British Royal Navy, or explorers sailing for the French or Spanish kings. One notable exception was the charts made by the Hudson Bay Company (HBC) employees. Their maps had little circulation outside the Company and were intended for internal use only. The HBC, chartered by the British government in 1670, had a strong hold on a vast area from Hudson Bay to the Rocky Mountains and north to the Arctic Ocean. Their maps were seldom published and generally were intended to provide details about the terrain and river systems in the Canadian interior. Their employees, such as Dease, Simpson, and Rae, made land surveys of immense areas of the Arctic coast and the near adjacent islands of the Canadian archipelago.

Through mapping and description, a government or commercial entity exerted their power over the land. Through their documents, charts, and narratives, the land of the imagination was transformed into a reality over which they ruled. The documents provided proof that a representative of the government had been there and established the power of the nation in that area. If they failed to map, they failed to control. The Arctic areas of Canada, for example, became a widely known reality under British control by first being drawn on paper and described in detail.[4]

Maps are deceptively simple things to view, and it is easy to believe they have their origin on the cartographer's table. The true origin of a map occurs much earlier. The map of a coastline is nothing more than a line on a piece of paper, showing nothing about the experience of being in a place. There is no indication on a map that a ship sailed for the first time into a new wilderness. To the person viewing a map, there is no hint that a few courageous men risked, or lost, their lives to produce a line on a chart. We rarely think of the labor that produced it—the journey to each point, the hardships in mapping those points. Historian of science, D. Graham Burnett, wrote: "I am particularly interested in how certain explorers—lightly equipped, underfinanced, more or less solitary—struck out into difficult environments, spent months and even years wandering in the interiors of continents far from home, and came back with maps, maps that ended up on negotiating tables in London, Paris, and Berlin."[5]

We seldom think of the surveying instruments used for measuring angles to a star, or the chronometer for calculating longitude—all the equipment needed to create the many points on a coast or stream so that a faithful representation can be drawn. Nor are we aware of the hours spent writing a narrative from notes taken during the survey—with a quill pen or a pencil. We enjoy reading the adventure story without pausing to reflect on the meticulous planning that went into an expedition—how to carry enough food and prepare for contingencies. Readers seldom consider the enormous costs in money, effort, and lives that resulted in some lines on a map, a written narrative, and a few illustrations. These high costs required kings or groups of wealthy men to underwrite such immense endeavors. But of course, the financial backers expected more than a map. They had expectations of financial gain from new resources, hopes of expanding empires, dreams of a new route to Asia, and simply the fame that came with important discoveries. A common motivation among kings was to claim land before a rival country could reach it.

Besides needing charts to help the next explorer find his way, the first owners of these charts had issues of power in mind—for confirmation or expansion of their dominion over the land. Their power is infused into the map.[6] In the earliest maps of North America, cartographers drew flags as emblems of the royalty claiming the land, hence showing the connection of power and land. When we read of expeditions through the nineteenth century, we are repeatedly struck by the audacity of the European belief that any land not ruled by a "civilized and Christian" monarch could be claimed, settled, and exploited. The concept of exploitation included taking resources, capturing hostages, and subjugating or enslaving a native population. These conditions were implicit powers embedded in the map. Who made the map ruled at will. Surveying and mapping an unknown land, whether done accurately or not, announced to the world that the distant land was now a known place ruled by a European king.

2

John Cabot Makes a Claim for England, 1497

THROUGH THE FIFTEENTH century, the winds of change were leading the king-
doms of western Europe toward the modern states. By the beginning of the sixteenth
century, Spain had become unified, the Tudors held power in England, and Henry
VII wanted the benefits enjoyed by Spain and Portugal. Some amazing voyages that
discovered new lands, new cultures, and new wealth in resources discredited the
world view of the Mediterranean as the center. The European economy was greatly
expanded by these new resources, and the centers of prosperity began a slow shift to
the north and west.[1] In this environment of change, Spain sent Columbus, and
shortly afterward, England sent John Cabot.

Though he probably never set foot on the mainland of North America, John
Cabot (ca. 1451–ca. 1499) certainly found and claimed a new land in 1497 that is
still called Newfoundland.[2] Cabot's mission was to find a route to Asia, and when he
rounded the south coast of the island of Newfoundland and beheld the open expanse
of ocean ahead of him, he returned to England. Reporting success to his king and
investors, he hoped to obtain funding for a second voyage designed for trade with
the people of Asia. Thinking that the lands they had encountered were some large
islands that could be easily bypassed, they needed only to find the way.

Considering that John Cabot presumed to claim much of North America for
the English, there is surprisingly little known about him. There are no portraits,
no descriptions of his appearance or personality, nor any letters or ships' logs
written by him. The site of his landfall on the island of Newfoundland is not
known for certain, and there is disagreement on almost every aspect of his first

voyage. The entire second voyage is unknown, except that he departed from Bristol in 1498 and was never seen again. Even at this early time, skilled mariners such as Cabot kept daily logs of locations, bearings, and landfalls throughout the voyage. Often a mate or other sailor on the crew also kept a private account. Yet none of Cabot's records exist today. The written material that does exist about Cabot comes from letters by his contemporaries interested in the voyage of discovery, particularly John Day of Bristol and Lorenzo Pasqualigo, a Venetian residing in London. John Day knew Cabot personally and presumably had access to his logs. In this way, we have come to know the latitudes he traveled and descriptions of his sightings. The information from John Day is in a letter to Columbus, whom he addressed as "the Grand Admiral."

Cabot's son, Sebastian, sailed as a boy with his father and provided accounts of the first voyage. Some aspects of his version are inconsistent and are now viewed as mostly self-aggrandizing, taking undeserved credit for discoveries and lacking credibility. Sebastian's account of the discoveries became the basis for many history books through the early twentieth century, but questions arose later as his accounts came under more scrutiny.

One certain fact is that in March 1476, John Cabot (Giovanni Caboto) was a newly declared citizen of the Republic of Venice and had been living in that city for fifteen years. Since estimates of his birth date point to 1451, he must have been a lad of ten years when he went to Venice with his family. Some historians think the family probably moved there from Genoa, John's birth place. Placing John Cabot in Genoa between 1451 and 1461 makes it likely that he knew Christopher Columbus, who was also born in Genoa in 1451. Perhaps they spent time together as boys at the pier, watching ships and talking to sailors. Cabot and Columbus both learned the seafaring trade by signing on with ships out of Genoa.

The next place of residence for John Cabot is believed to have been in Valencia, Spain, between 1490 and 1493. If this is correct, Cabot might have been aware that Columbus was sailing to the west in search of Asia, and could have met with Columbus in 1493 as the famous Admiral passed through Valencia en route to Barcelona to report to the king after his first voyage. In any event, by this time Cabot was inspired to become involved in the search for routes to Asia. He began to seek support for such a venture in Portugal and Spain with no result. Next he turned to England to promote his idea that a northern route to Asia would be much shorter. His navigational experience had taught him that a transatlantic passage at high latitude would be shorter than at the lower latitudes traversed by Columbus.

England, like most of Europe, had developed a taste for spices, particularly pepper, nutmeg, cloves, and other exotic wares from the East, but because of their distance from sources, the English had to pay premium prices for these luxuries. King Henry

VII saw an opportunity to seize a piece of the action from Spain and Portugal. He agreed to support a voyage commanded by John Cabot to search for a shorter northern route for the spices of South Asia—at no cost to the crown. Cabot had offered to explore at his own expense under the King's sanction. As it turned out, the merchants of Bristol funded the venture.

In the fifteenth century, Bristol was already a major center in England for shipbuilding, manufacturing, and international trade—second only to London. Many wealthy Bristol merchants were involved in import and export, and they had great interest in enhancing Bristol's position as a major player in world trade. When John Cabot arrived in England in 1495, it was Bristol where he and his wife Mattea and their three sons settled, and it was the Bristol merchants whose support he sought for his voyage. Perhaps Cabot met some of these merchants while he was still in Spain, where he already had a solid reputation as an able navigator, and they may have encouraged him to come to England to apply for royal letters patent granting him permission to explore and claim new lands. Bristol merchants were especially interested in finding a fishing area to compensate for their exclusion from the Iceland cod fishery by the Hanseatic League, which maintained a trade monopoly over the area.

In March 1496, Henry VII granted letters patent to John Cabot and his sons, Lewes, Sebastian, and Santius, with authority to sail with five ships under Henry's banner to all parts of the east, west, and north to "seek and find lands of the infidels and heathens," which were not known to Christians before this time. Cabot and his sons could govern these newly claimed lands in the King's name and could profit from the produce of these lands, paying 20 percent to the crown. Anyone wanting to visit these lands would need the permission of the Cabots. Such letters patent were required authorization to establish the validity of discovered lands. Without such authorization, the discoveries would not be recognized by anyone.

International recognition of Cabot's discoveries was doubtful from the beginning. As soon as the importance of Columbus's discoveries was known, Pope Alexander VI issued a Papal Bull, the Treaty of Tordesillas. This treaty, first issued in 1493 and modified in 1494, divided exploration and exploitation rights in new lands between Spain and Portugal by a north–south line of demarcation 370 leagues (about 1,275 statute miles) west of the Cape Verde Islands in the Atlantic Ocean. This line gave the lion's share of both new continents to Spain, leaving Portugal with only the eastern projection of Brazil. However, Portugal was able to push westward into the interior of Brazil with no resistance from Spain. Later the line was extended around the world, giving Portugal a chance for claims in Africa and Asia.

Spain and Portugal expected England and all others to stay out of their global land grab. On that basis, the Spanish king protested to England when Cabot ventured into the new lands. Henry VII was unimpressed and ignored the protest.

THE FIRST VOYAGE

Cabot's first attempt in 1496 to find a northwest route to Asia had been aborted, and the ship turned back. John Day's letter, the primary source of details about Cabot's voyage, says that his crew "confused him, he went ill provisioned and encountered contrary winds and decided to return."[3] The most likely explanation of the failed voyage is that a disagreement arose between Cabot and the crew regarding routes to the new lands. The crew were from Bristol and probably knew about a route via Iceland, Greenland, and Baffin Island to fisheries near Vinland. When they saw Cabot heading straight west from England, they must have convinced him that they would be going too far south. The disagreement may have become an open rebellion, forcing Cabot's return to Bristol. The Bristol financial backers still had faith in Cabot and quickly put together another voyage without reporting to the king, lest he lose his commitment to the project.

Although the letters patent authorized five ships, Cabot had support for only one: a small, fifty-ton caravel. By May 1497, Cabot was ready to begin the new voyage—probably with a new crew. Cabot had a new caravel built in Bristol and christened it *Mathew*.

Cabot left Bristol on May 20, 1497, with his young son Sebastian and a crew of eighteen men. He sailed northwesterly to a coastal prominence, Dursey Head, on the west coast of Ireland, then turned west. This was the typical method of sailing, called latitude sailing: sailing north or south until the desired latitude was reached, then sailing due west along that course. The chosen latitude was the one that he expected would take him to the northern coast of China. To stay on a westerly course along the selected latitude, they had to sight a known star or the sun using an angle measuring instrument like the astrolabe or quadrant.

Dursey Head, Ireland, is at latitude 51° 33′ N, so Cabot kept to a course that would maintain his guide stars at that angle above the horizon. If the angle was less, they knew they were drifting south, and if larger, they were drifting north. Had it been possible to keep that course all the way to Asia, they would have hit the southern tip of the Kamchatka Peninsula. The North American continent, however, intervened. Cabot had a compass, an astrolabe, perhaps a quadrant, and a traverse board for recording positions. These instruments allowed him to deviate from the chosen latitude and still know his location. More detail on instruments is provided in Chapter five.

The *Mathew* reached Newfoundland late in the afternoon on June 24, 1497. Cabot apparently provided no information about his landfall, so the exact location is unknown. However, the most probable site is at the northeast portion of the northern peninsula of Newfoundland Island (figure 2.1). Cabot saw a large island to the north of where he landed and named it St. John's Island in honor of John the Baptist, who is commemorated on June 24th. French explorers later renamed the island Belle Isle, which is the current name. The *Mathew* had sailed from Bristol to Newfoundland in thirty-five days. The voyage from Dursey Head, Ireland, the last point of land in the British Isles, to Newfoundland took only thirty-two days, a record speed that held for almost a century.[4]

Cabot's probable landfall was remarkably close to L'Anse aux Meadows, the site of Leif Ericson's Viking settlement established nearly five hundred years earlier in 1001. English sailors in the fifteenth century had no knowledge of that settlement, so Cabot would have assumed that he was the first European to visit the site. By a further strange coincidence, thirty-seven years later, the French explorer Jacques Cartier made his North American landfall in almost the same spot.

At that latitude in June, there would have been a long twilight, providing plenty of time to search the coast for a safe anchorage before nightfall. Cabot began sailing south down the outer coast of Newfoundland Island. The inner coast, Belle Isle Strait, would no doubt have been blocked by ice in June in the fifteenth century when the Little Ice Age was in bloom. It is not known which inlet he chose for a harbor, but a likely choice appeared less than 6 miles down the coast from the tip of the peninsula. This harbor, now called Griquet Harbor, is well protected and calm, and Cabot would have been able to make accurate latitude readings, which he found to be 51°33′, the same as Dursey Head, Ireland. Skilled navigation had brought him to this spot.

Some historians believe John Cabot's landfall to be Cape Breton, Nova Scotia. Samuel E. Morison, the respected authority on Columbus and other early ocean explorers, deems this an error. According to Morison, Cabot measured a latitude of 51°33′ N at his landfall, and the nearest part of Cape Breton is 47°02′ N.[5] Such a large error (4.5°) in navigation and reading latitude is unlikely for an experienced mariner such as Cabot. Cape Breton is more than 300 miles south of Cabot's reading. Cabot would have taken meridional (noon) altitude of the sun and night sighting of Polaris, either while anchored in harbor or perhaps on land, to determine latitude. To suggest such a major error underestimates the navigational skills of the best sailors of that day. Cape Bauld (lat 51°37′ N) was probably the first land sighted, and Griquet Harbor (lat 51°33′ N) would be the first likely anchorage.

At this harbor, Cabot would have gone ashore for a stable place to measure latitude and to perform a brief ceremony taking possession of the land for King Henry

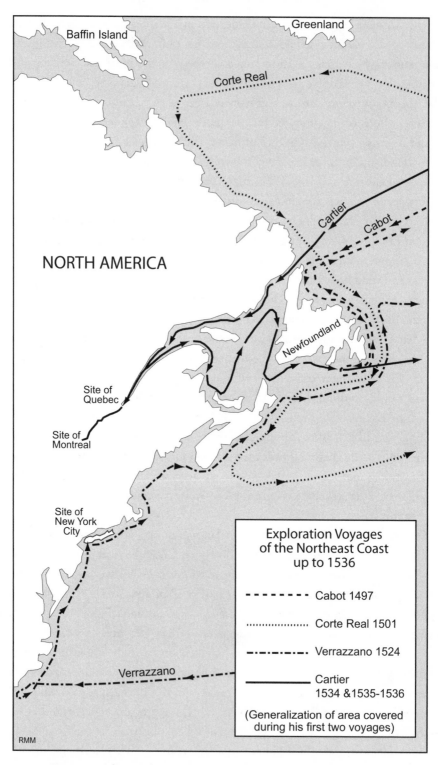

Baffin Island

Greenland

Corte Real

NORTH AMERICA

Cartier

Cabot

Newfoundland

Site of
Quebec

Site of
Montreal

Site of
New York
City

Verrazzano

**Exploration Voyages
of the Northeast Coast
up to 1536**

- - - - - - - Cabot 1497

.... Corte Real 1501

-·-·-·-· Verrazzano 1524

———— Cartier
1534 &1535-1536

(Generalization of area covered
during his first two voyages)

RMM

FIGURE 2.1 Four routes of sixteenth century voyages to North America.

VII. The ceremony included implanting the flag of St. George, the emblem of England. Such possession usually implied inclusion of all adjacent land that had not previously been claimed, and so could be interpreted to mean the entire continent.

Cabot wrote that he saw signs of habitation, such as snares and fishing nets, but no people. He "dared not go inland farther than an arrow shot" for fear of hostile natives. This was the only time Cabot recorded going ashore.[6] Four years after Cabot's landing, the Portuguese explorer, Gaspar Corte Real, came to this site, abducted some Beothuk Indians, and took them back to Portugal. The Indians had an Italian-made sword and earrings made in Venice, which must have come from Cabot.

Cabot wrote of dropping weighted baskets in the Grand Banks area and pulling up loads of codfish. He reported large trees suitable for ship masts and climate that was mild enough for growing silk and logwood, which was a valuable source for red dye.[7] The report of codfish elicited a quick response from English fishermen, who soon began making the journey to harvest cod from the Grand Banks off Newfoundland. Although the fishermen had no interest in exploration, they, along with whalers, soon became an important source of information about the conditions of seas, harbors, and passages off eastern North America. Explorers planning a western voyage learned to query the fishermen for valuable information about what to expect in those foreign waters.

Cabot sailed down the east coast of Newfoundland Island to the south end of the island. There he looked westward and saw no land on the horizon. Thinking he had found the sought-for passage to Asia, he turned back. He retraced his route northward along the east side of Newfoundland, completing his mapping of the coast. He then turned eastward to England on July 20, 1497, along the same latitude as his outward voyage.

According to the contemporary account by John Day, Cabot's crew persuaded him that their course should be more southeasterly. This deviation resulted in their making first landfall on the coast of Brittany, France, on August 4, 1497—only fourteen days of traveling with the westerly winds. Another two days took them back to their home port of Bristol. There was no reason a seasoned navigator like Cabot should be swayed to change course when the outward voyage had worked so well. A similar change had apparently occurred a year earlier with the false start to the first attempted voyage. The ease with which Cabot was influenced by his crew probably indicates that they were all locals claiming to know more about sailing in the area, and Cabot was still regarded as a foreigner. The crew again proved to be mistaken; Cabot would have done better to rely on his experience and instincts.

They sailed the *Mathew* up the Avon to Bristol, perhaps being towed part of the way. After greetings in Bristol, Cabot left for London, and, on August 10, he arrived in Westminster to report to King Henry. The king paid Cabot £10 and in early 1498 granted him an annual pension of £20 (equivalent to about $90,000 today) for the discovery of Newfoundland, and the two discussed plans for another voyage.[8] The king promised armed ships and prisoners would accompany Cabot, suggesting that he intended to establish a colony and trading. Ship's captains in that time earned about £8.5 per year, so the £10 bonus plus £20 annually was a handsome amount. The king made it clear that the pension was to be paid from the customs income at the Bristol port. It was easy for the king to be generous, as it cost him nothing. The Bristol customs office apparently was a bit short on funds, for Cabot did not receive any payment for two months. The last recorded payment was one year later when payments to John Cabot stopped appearing in the ledgers.

Cabot returned to Bristol where he was a celebrity and people were eager to hear of his discoveries. It is hard to imagine the enormity of learning about new lands that were totally unknown to them. Cabot was now called "Grand Admiral" with great honor, and people gathered wherever he went.

Spain accepted England's claim in the New World due to political expediencies, though with some reluctance. The Spanish king, Ferdinand V, was bothered by Henry's discovery in "Spanish" territories but did not want to protest too strongly as he needed England's support against the French in Italy. Henry VII wanted to stay on Ferdinand's good side because he hoped to wed his son, Arthur, to Ferdinand's daughter, Catherine of Aragon, and possibly gain control of the Spanish throne at some future time.

As early as August 1497, Lorenzo Pasqualigo, writing to relatives in Venice, said that Cabot claimed that his discovery was China and was calling himself the Great Admiral. Pasqualigo also wrote that Cabot was boasting that the king would give him a fleet for his next voyage and criminals to found a colony on lands he would discover. John Cabot was now preparing to tread openly in Spain's designated sphere. In February 1498, Henry VII provided new letters patent granting Cabot permission to represent England for a second voyage. Further, the king granted him the right to take six ships of two hundred tons or less. He was directed to sail beyond his previous voyage until he reached the island of Cipango (Japan), which they considered to be the source of the coveted spices. There, Cabot was to set up a colony as a trade center for goods to be shipped to England.

THE SECOND ATTEMPT

As it turned out, the king provided only one ship and the merchants of Bristol outfitted four more. Cabot's plan for 1498 was to begin at Newfoundland and sail west and south until he came to a passage. He had again been given the right, according to his letters patent, to "subdue, occupy and possess isles, countries, regions, or provinces of the heathen and infidels unknown to Christians, and to enjoy their fruits, profits, and commodities." He would avoid claiming land that might infringe on Columbus's discoveries in the Caribbean, but would just follow the coast of the mainland. No one was yet aware of the vast extent of the land mass that blocked their way to Asia.

The five ships left Bristol in early May 1498, outfitted with a year's provisions, though the trip was expected to be much shorter. They were loaded with trade items such as cloth caps, laces, and other trifles. The merchants of Bristol had little idea what trade items might be most valued by the unknown Asian sellers of spices.

One ship suffered damage from a storm and had to put into an Irish port, but the other four proceeded—and were never seen or heard from again. Cabot did not achieve his objectives, and he probably never knew the significance of his discoveries, but he was the first of a long English presence in North America. In Bristol, his pension was paid in 1498 and 1499, but the lease of his house was terminated in 1499, leaving Mattea and her children in the care of the city authorities. By 1512, John Cabot was declared lost at sea. After Cabot's second voyage, there was a lull in English exploration because of the unwillingness of Henry VIII (reign 1509–1547) to continue funding unprofitable voyages. Lack of interest in exploration continued until the reign of Elizabeth I (reign 1558–1603), who sanctioned and helped fund the next round of exploration.

Cabot established that a substantial land mass did exist within reasonable sailing distance from Europe and laid the way for the Bristol fishery in the Grand Banks (the first catches were brought home in 1502 and 1504). Cabot is also credited with introducing the logbook for mariners to keep track of their routes. This innovation made reconstruction of voyages much easier and more accurate for mapmakers.

His achievements were commemorated at the 400th anniversary of Cabot's first voyage, by the building of the Cabot Memorial Tower in Bristol in 1897. A century later, Cabot's voyage was celebrated again by the construction of a replica of the *Mathew*, which successfully crossed the Atlantic to Newfoundland, touring the east coast of Canada and the United States before returning to its permanent home in Bristol harbor.

John Cabot's son, Sebastian, was a puzzling character. He probably made no contribution to the map of North America, although he claimed to have been the discoverer in 1496, the year before John's voyage. Sebastian's birth date is uncertain, and plausible estimates of 1482 would make him only 14 years old at that time. Sebastian continued living in Bristol, earning a living as a merchant. In 1505, he was the principal beneficiary of a new royal grant, which also awarded him a pension of £10 a year in consideration of his services.[9] These services and the work that became his profession were probably performed in Bristol rather than at sea: cartography and the study of navigation.

Sebastian may have sailed in 1508, at age 26, with two ships as far as 55° N latitude but was forced to turn back because of ice. He claimed to have discovered the entrance to the bay that later became Hudson Bay. The only well-documented voyage made by Sebastian was in 1525–1528 for the king of Spain with the object of sailing around the world, repeating Magellan's feat. This voyage also was aborted, although it is known that he managed to receive payment for his efforts.

On a painting of Sebastian, which he commissioned himself, is an inscription in Latin: "Sebastian Cabot, Englishman, son of John Cabot, Venetian knight, first discoverer of Newfoundland under Henry VII, King of England." The statement could be read to mean either that he claimed himself to be the first discoverer or gave credit to his father. This is the only known reference to John Cabot as a knight. Perhaps that was an honor bestowed by Sebastian, who was adept at altering history.

There is a record of Sebastian's pension payments stopping in 1557, probably the year of his death. One contemporary writer said that Sebastian claimed on his deathbed to be able to know longitude by divine revelation. He seems to have built a myth of super powers and capabilities around himself.

For about twenty-five years after John Cabot's voyage of 1497, only cod fishermen sailed to the Newfoundland area, creating a lull in the mapping of North America. During that time, Central and South America continued to be developed, with their riches going to Spain and Portugal. These southern areas had such wealth that the Spanish and Portuguese gave little thought to looking elsewhere. The north offered only tall trees, codfish, and a lot of ice. America had been found by accident, and much of the subsequent effort was directed toward finding a way around or through it to complete the voyage to Asia. Little was done to expand the limits of the map of the North American continent, except for a little known voyage by Gaspar Corte Real.

Gaspar Corte Real made two voyages into the North Atlantic— to Greenland and Newfoundland—with the authority of the king of Portugal in 1500 and 1501. In the second voyage, only two of Gaspar's ships returned, this time with fifty-seven Indians of the Beothuk tribe that they had kidnapped. These Indians possessed the Venetian sword presumed to have come from Cabot's first or second expedition. The third ship, with Gaspar himself, was lost at sea. In 1502, Gaspar's brother Miguel (ca. 1448–1502) set sail with three ships to find his brother. Miguel's ship was also lost, and no new information of North America was gained from the expedition.

A world map made from the information of Gaspar's voyages was made in 1502 by Alberto Cantino for the Duke of Ferrara and still exists in the Estense Library in Modena, Italy. The Cantino Mappemonde of 1502 was the first to include information from Portuguese voyages to the New World. The map ignores any English claims, even though the Portuguese certainly knew about them and Gaspar had found traces of Cabot's visit to Newfoundland. The map shows the south edge of

Greenland labeled as a part of Asia, and shows the Portuguese king's flag on both Greenland and Newfoundland. After all, they presumed that the pope had given them the right to these areas.

Although the cartographer of the Cantino map is anonymous, Alberto Cantino, Duke of Ferrara, acquired the map and it became known by his name. The caption of the map shows Newfoundland 15° east of its actual location, making a portion of it east of the Tordesillas line and therefore legitimate Portuguese territory at that time. This discrepancy could have been intentional to expand their area, or it could be accounted for by the uncertainty of the longitudes of their claims. At any rate, they did nothing to maintain the claim through settlement or even by sailing again to the area. Portuguese and English fishermen, on the other hand, continued to fish the waters near Labrador.

3

Giovanni da Verrazzano Maps an Ocean of His Imagination, 1524

EXPLORATION AND LAND claiming spread through the Western Hemisphere from the time Columbus began his voyages in 1492. In 1521, Spain began a three hundred-year rule of Mexico, and Magellan was making the first circumnavigation of the world, claiming whatever he encountered. That expedition returned to Spain with one of five ships and only 18 of the original 239 men that left in 1519. Magellan himself died during an disastrous offensive with natives in the Philippines, and Juan Sebastian del Cano assumed command of the voyage. The account of this voyage, published in 1523, stimulated a great interest in more exploration for routes to Asia.

Up until 1520, the efforts of the Spanish and Portuguese to find a passage through North or South America were defeated. They searched the southern part of the east coast of the Americas from Florida to Patagonia and found no strait connecting the Atlantic to the Pacific, except for those Magellan had sailed around the horn of South America. However, vast areas to the north—from Florida to Greenland—still held possibilities. John Cabot had claimed land at Newfoundland but returned home without searching for a westward passage. The east of North America remained unexplored, and everyone concerned believed that a sea route must surely exist somewhere along such an expanse of coastline.

Merchants and bankers in France saw a profitable advantage if they could buy the silks of China without a long voyage south through Spanish-dominated waters. They gained the eager approval of Francis I, king of France, and in 1523 commissioned Giovanni da Verrazzano to explore the east coast of North America from Florida to

Newfoundland. France had at last entered the exploration of the New World. Verrazzano acquired a ship, *La Dauphine*, on loan from the French navy, and the merchants provided a second ship, *La Normande*. These ships were approximately one hundred tons in size with crews of fifty to sixty men. Verrazzano's brother, Girolamo, a mapmaker, also went on the voyage.

Born to a wealthy family near Florence, Giovanni da Verrazzano (1485–1528) had the best education available, with well-developed skills in mathematics. In 1506, he relocated to Dieppe, France, at the age of 21, and there he improved his navigation skills. He apparently gained experience on ships sailing in the Mediterranean to the Middle East. In 1508, he may have joined a voyage to Newfoundland on a French ship.

By this time, Verrazzano excelled at navigation and doubtless kept detailed records of his positions, as any mariner should. Although the logbooks are missing, the maps resulting from his voyage are accurate in terms of latitude, which he probably measured with great skill using an astrolabe or quadrant. Longitude, of course, was not accurate but was close enough to make the eastern coastline of North America recognizable. The cross-staff had been recently invented at the time of Verrazzano's voyage, but it is unknown if he used it. The quality of his latitude data hints that he may have had this more accurate device for measuring vertical angles. Verrazzano wrote a detailed account of his voyage in a lengthy letter to Francis I.[1]

In January 1524, Verrazzano sailed southward from France to the Portuguese-owned Madeira Islands. As it turned out, *La Normande* had to turn back and *La Dauphine* continued alone. From the Madeiras, he turned west on a course carefully planned to avoid contact with Spanish warships. French pirates had been plundering Spanish ships loaded with wealth from the New World, and Spain had become hostile toward any French ships traveling in those low latitudes. He steered west along 32°30′ N latitude, a course that would have taken him to the coast of North America about thirty miles south of Charleston, South Carolina. See figure 2.1 for a map of Verrazzano's route.

For the first few weeks of the voyage, *La Dauphine* enjoyed the steady push of northeasterly trade winds. In late February, they sailed into a severe storm and had to veer north to escape the winds and save the ship. Verrazzano then set a new course at 34° N latitude and eventually made landfall at Cape Fear, North Carolina. From this landfall, he turned south to try to reach the destination originally set. However, Verrazzano sailed south for only about 120 miles before turning back north to be certain of avoiding the Spanish. His turning point was possibly near the present site of Charleston. On reaching Cape Fear again, Verrazzano dropped anchor and sent some men ashore to meet the natives who had come out to see them. The Indians treated them in a friendly way and offered the crew some food. Verrazzano described the Indians as strong and well built, with appearance and behavior somewhat like the Chinese. This conclusion must have arisen from his familiarity with the writings of Marco Polo, along with some wishful thinking that may have prompted him to see Chinese characteristics in the Indians.

Moving farther up the coast, Verrazzano noted sweet aromas coming from the land and entered the names *Selva di Lauri* (Forest of Laurels) and *Campo di Cedri* (Field of Cedars) for the area. These locations are not identifiable, but laurel and cedar still grow along the Georgia and South Carolina coast.

At his next anchor point up the coast, Verrazzano sent twenty-five men toward shore in a boat. Because no harbor existed there, the surf was so large that the men feared their boat would capsize. Indians appeared showing signs of friendliness and urged the men to come ashore, but the men would not take the risk. The crewmen had brought gifts for the Indians, so one man swam ashore for the purpose of meeting the Indians and giving them the gifts. As he tried to return to the ship, he found the surf impassable and was badly beaten about and bruised in the effort. The Indians came to help him and carried him onto the shore and made a fire to warm him. When his shipmates watched the Indians stripping off the man's clothing and building a fire, they felt sure the poor man would be roasted and eaten. After a short time, the man recovered his strength, put on his clothes, and indicated he would return to the boat. The Indians showed him great affection with many hugs, and the man again entered the surf. This time he made it back to the boat and to the ship.

By this time, the *Dauphine* was sailing in the waters off the long barrier islands, the Outer Banks that form the coastline of much of North Carolina. These waters are famously dangerous for two reasons. Here, the cool Labrador Current flows south and converges with the warm Gulf Stream, creating an area of severe turbulence. In addition, shallow sandbars create dangerous shoals for several miles out from the shore. Because of these hazards, Cape Hatteras earned the reputation as a "graveyard of ships." Verrazzano must have perceived the problems for navigation in the area, as he seems to have sailed well away from the coast.

Verrazzano's practice of keeping a good distance from shore may have caused a great error that he made in this location. From his deck on the *Dauphine*, Verrazzano could see the water of Pamlico Sound on the other side of the barrier island, but he could see no land beyond. He jumped to the conclusion that the barrier island was merely an isthmus of land 200 miles long and only 1 mile wide, and that Pamlico Sound was the Pacific Ocean. He named this supposed isthmus Verrazzania.

It should be said in his defense that from a position seaward of the barrier islands, the low-lying mainland shore is not visible.[2] However, Verrazzano's fault lay in not sending a boat through one of the openings in the barrier to investigate further. One could argue that the openings in the barrier bars may not have existed in 1524.

Although the position of the openings may shift over time, the openings must exist as long as rivers continue flowing into Pamlico Sound. Fresh water from rivers must break through to the sea at some places along the 200-mile length. Either he could not see the openings or he deemed them too narrow for passage. Nevertheless, for a century afterward, maps showed the Sea of Verrazzano splitting the North American continent. This imagined sea, shown in figure 3.1, has become one of the more famous errors in early mapmaking.

At some point along this part of the coast, Verrazzano's men went ashore and kidnapped a child from a woman.

By searching around we discovered in the grass a very old woman and a young girl of about eighteen or twenty, who had concealed themselves; the old woman carried two infants on her shoulders, and behind her neck a little boy eight years of age; when we came up to them they began to shriek and make signs to the men who had fled to the woods. We gave them a part of our provisions,

FIGURE 3.1 Giovanni Verrazzano imagined that Pamlico Sound was an ocean offering a route to the Orient. Cartographers called it the Sea of Verrazzano.

which they accepted with delight, but the girl would not touch any; everything we offered to her being thrown down in great anger. We took the little boy from the old woman to carry with us back to France, and would have taken the girl also, who was very beautiful and very tall, but it was impossible because of the loud shrieks she uttered as we attempted to lead her away; having to pass some woods and being far from the ship, we determined to leave her and take the boy only.[3]

Such behavior, it seems, was standard for sixteenth-century Europeans upon their first sighting of Indians. Nothing is known of the fate of the child, but in other known instances of kidnapping, the victim did not live long in the new country. Verrazzano went to some length in describing the Indians' way of living: describing how they made dugout canoes from a single tree trunk; telling that they ate fish, fowl, and beans; and explaining that they wore leaves for clothing. He named this part of the coast Arcadia after the idyllic place in Greek literature. Later cartographers moved the name Arcadia (becoming Acadia) farther north to include parts of present day Nova Scotia and Maine.

As Verrazzano continued north along the coast, he may have been in some way less attentive to the land, possibly due to weather conditions. As a result, he made a glaring omission in the resulting map by completely missing both the Chesapeake Bay and Delaware Bay. Either of these large inlets would have attracted the attention of anyone searching for a westward passage. Another explanation is that his caution regarding shoals near the shore caused him to sail too far seaward to see much detail of the land. A mariner's first rule is to avoid grounding his ship. In view of the subsequent history of shipwrecks along the Outer Banks, his caution may have been justified. Verrazzano fortunately sailed the east coast in the spring without encountering a sudden major storm, another common cause of shipwrecks in the area.

Verrazzano's next discovery still stands as his most notable: New York Bay. He and his crew became the first Europeans to see this beautiful safe haven. He anchored the *Dauphine* in the bay's narrow entrance, and it is especially apt that they are now named Verrazzano Narrows and are spanned by the Verrazzano-Narrows Bridge.

He explored the bay briefly from the ship's boat, and reported seeing many Indians onshore. Unfavorable winds prevented him from going ashore to meet them. After only one day in the bay, which he named Santa Margarita, he weighed anchor and proceeded up the coast. He missed exploring the great river flowing into the bay, leaving that to Henry Hudson eighty-five years later.

Farther along, Verrazzano described and mapped an island that he described as being about the size of the Greek island of Rhodes. He named it Luisa, and it might have been the island now known as Block Island off the coast of Rhode Island.

Alternatively, it could have been the island that the well-educated founder Roger Williams named Rhode Island, now the site of Newport in Narragansett Bay. Williams might well have read Hakluyt's translation of Verrazzano's account and applied the name that also became the name of the Rhode Island colony.

Verrazzano temporarily set aside his cautious practice of always dropping anchor offshore in open water and sailed directly into Narragansett Bay. He noted that the islands near the entrance would provide an excellent location for coastal fortifications, which were in fact used during the Revolutionary War. Indians in canoes guided him into a safe harbor, probably the present Newport Harbor, and made quite a favorable impression on Verrazzano. He described them as extremely handsome, strong, and well proportioned, with well-trimmed hair. They wore elaborate headdress, which he compared to those of the women of Egypt and Syria, and described that some of them wore skins of deer and lynx. He noticed that the Indians (the Wampanoag tribe) had wrought copper plates that they valued highly and used for trade. This site must have been much more attractive than New York Bay, as Verrazzano stayed there for fifteen days.

Continuing up the coast, the *Dauphine* passed Cape Cod, which Verrazzano identified only as an outstanding promontory. Past Massachusetts Bay to the coast of Maine, Verrazzano described the land as fair and open with high mountains inland, probably the White Mountains. The Indians he encountered there (the Abnaki tribe) exhibited no sign of friendliness. Whenever his crew tried to row ashore, the Indians began shooting arrows at them. Later the Indians agreed to trade for knives and fishhooks by lowering baskets from a cliff to the boat below. It appears that the Abnakis may have encountered, or heard about, the kidnapping tendencies of Europeans. As they departed, the French crew saw the Indians making uncouth signs and displaying disdain. Essentially, they were "mooned" by the Indians from the cliff tops.[4] Verrazzano, in a state of disgust, decided to name this stretch of coast *Terra Onde di Mala Gente* (Land of Bad People).

Verrazzano described many islands and fine harbors along the Maine coast. He noted the springtime beauty of the region with its blooming flowers and many tall, straight trees suitable for making masts. The *Dauphine* soon ran into easterly winds, and Verrazzano steered farther offshore as they continued up the coast. As a result, he again missed some prominent features, particularly the Bay of Fundy and most of Nova Scotia.

His next encounter with land was the area previously visited by John Cabot—Newfoundland. There Verrazzano restored his supply of fresh water and wood and began the return to France. They had surveyed some 2,200 miles of previously unmapped coastline. He missed a few important features, but in one grand sweep, Verrazzano filled in the largest single gap in European charts of North America. Although he believed the great Pacific Ocean lay just beyond the barrier islands of

North Carolina, he definitely showed that no passage existed along the east coast of North America as far as the Island of Newfoundland. Many subsequent explorers improved on Verrazzano's map of North America and corrected his errors, but the conclusion was obvious: the only remaining unexplored area that might contain a passage was the Arctic. Logic assured European mapmakers of the sixteenth century that the passage in the far south must be matched by another in the far north. It was simply a matter of balance and symmetry. As the southern passages were dominated by the Spanish and Portuguese, the northern European seafaring nations desperately wanted to find a northern passage. They spent the next 380 years trying to batter the barriers of ice blocking the Northwest Passage.

Despite his accomplishment, Verrazzano was unable to generate support for another voyage because of a war that required the French king's attention, treasury, and ships. Verrazzano eventually found support elsewhere for a commercial enterprise with the promise of a share of profits. That voyage left Dieppe, France, and went to Brazil in 1527, bringing home a load of brazilwood (*Caesalpinia echinata*), much valued in Europe for making red dyes for textiles. This may have made him some money, but did not equal the personal satisfaction of looking for a water passage to the Pacific.

The investors gave Verrazzano another chance in 1528 to explore the coast of Central America, searching for a possible strait and bringing home another load of brazilwood. In the island chain of the Lesser Antilles, Verrazzano anchored offshore of one of the islands while he and six crewmen approached the shore in a boat. Once Verrazzano reached the shore, many Carib Indians attacked him and his men, cut up their bodies, and began eating them on the spot in full view of the men in the ship. Girolamo, the brother, and the shocked crewmen watched in horror as the sand turned red with blood, but they were too far out to help even with their guns. There was nothing for them to do but return to France.

By showing there was no water passage to the Pacific, Verrazzano had produced a negative result, which is typically of interest to no one. What Verrazzano accomplished, however, was to determine that North America was a vast continent, and that it possibly contained many resources of value. He also produced the data needed for the first rudimentary mapping of the entire east coast, from the Carolinas to Newfoundland, complete with numerous errors and some glaring omissions.

THE SPANISH ARRIVE ON THE WEST COAST

Less than twenty years after Verrazzano completed his traverse of the east coast of North America, two Spanish expeditions pushed northward along the west coast resulting in a corresponding expansion of the mapped coast of North America.

Their objective was to find the nonexistent Strait of Anian that an imaginative cartographer had created.[5] The first expedition was led by Juan Cabrillo, who in 1542, beginning in Mexico, mapped as far as San Diego Bay. His voyage continued north, but unfavorable winds forced him to stay well away from the coast, adding little to the map of North America. He turned back a short distance past San Francisco Bay, which he missed entirely. Bartolomé Ferrelo, who had accompanied Cabrillo, returned the next year and mapped 900 miles of coast up to Cape Blanco in Oregon. The Spanish discontinued further exploration up the coast, and for the next 235 years, Cape Blanco remained the northernmost point on the map of the west coast until Captain Cook extended the traverse in 1778. Beyond the wonderful bay at San Diego, the Spanish had found little to interest them in northern California.

The English privateer, Sir Francis Drake, arrived in the area in 1587 during his voyage around the world.[6] He was well known among the Spanish as a pirate, and his many raids on their ships earned him a reputation as a swashbuckler in England. His exploits were not only accepted but were also sanctioned and welcomed by the queen, who realized great financial benefits. Drake recorded no new coastline, but he went ashore in a bay, perhaps San Francisco Bay, and became acquainted with the Miwok Indians, whom Drake claimed "invited" him to claim the land for England. The precise location of the port was carefully guarded to keep it secret from the Spaniards. Invited or not, he claimed it and gave it the name New Albion. This bold presumption, of course, irked the Spanish and prompted them to establish themselves more firmly in California. In the end, Drake's claim had no lasting effect.

4

Jacques Cartier Gives France a Prize, 1534, 1535, 1541

BY 1532, THE golden lure of the New World had captivated all the maritime nations of Europe facing the Atlantic seaboard. France had already entered the land rush with Verrazzano, but despite his resulting map identifying broad areas of North America with the names *Francesa* and *Nova Gallia*, France did nothing to expand, settle, or otherwise capitalize on Verrazzano's discoveries. Their interest later revived and became focused on the exploration and discoveries of Jacques Cartier (1491–1557).

Cartier had an influential relative with a connection to the abbot of Mont-Saint-Michel and bishop of Lisieux. The abbot presented Cartier to King Francis I as the master mariner who could discover lands for France in the New World. Although this was only four years after Verrazzano's 1524 voyage, the king enthusiastically agreed to hire Cartier and to underwrite his exploration with the help of the bishop. Before anything could happen, the king had to get approval from Pope Clement VII to make exception to the earlier papal bull, the Treaty of Tordesillas, made by Pope Alexander VI dividing the undiscovered world between Spain and Portugal. The pope responded in the king's favor by declaring that the Treaty of Tordesillas applied only to lands already discovered by the Spanish and Portuguese, but not to new lands.[1] This interpretation of Pope Alexander's earlier decision put most of North America up for grabs.

The waters off the east coast of Canada had become familiar to fishermen soon after John Cabot reported masses of cod on the Grand Banks. But fishermen do not make maps, nor do they claim land for the king. For a nation to achieve ownership, someone had to go in the name of, and with sanction of, the king, and only such men produced maps of their discoveries.

Little is known of Cartier's background except that he was born in 1491 to a family of good standing in Saint-Malo. His family had a tradition of seafaring, and Jacques married well into a family of shipowners. By 1519, Cartier was designated a master pilot, which suggests that he had probably made voyages in the Atlantic, possibly to Brazil, prior to his voyages of discovery.[2] His skill as a mariner is attested by his three voyages of discovery for France to the uncharted waters of eastern Canada made without losing a ship or having a serious accident, and without losing any seamen except to ordinary illnesses. He had earned the respect of mariners, and for his first voyage of exploration to the New World, he held the title *Capitaine et Pilote pour le Roi*. Cartier was already a qualified pilot, but the title of *Capitaine* was added by command of the king.[3]

FIRST VOYAGE, 1534

Cartier began his first voyage for the king of France in 1534 at the age of forty-three, with instructions to find a passage to China and to discover precious metals. Cartier set sail from St. Malo on the 20th of April with two ships and crews of sixty-one men per ship. Twenty days later, he reached Cape Bonavista, Newfoundland, the standard destination of French fishermen. In mid-May ice was still a problem for navigation near Newfoundland, and Cartier had to move southward to find a safe harbor where they waited for ten days.

Coming out of the safe harbor after the ice had dispersed, Cartier sailed to Cape Bauld at the north end of Newfoundland—coincidentally the same spot where John Cabot and Leif Ericsson had made landfall. After a delay of several days waiting for the ice to clear, he sailed across the narrow Belle Isle Strait to Chateau Bay on the coast of Labrador, where he recorded seeing cranberries and an abundance of mosquitoes and flies. Here, Cartier was still in territory familiar to French fishermen. He then turned southwest to sail along the rugged western coast of Belle Isle Strait. This coast has many sailing hazards, such as offshore islets and submerged rocks, and the high opacity of the water required making constant soundings of depth to prevent running aground. All this vital information ultimately became part of the map and narrative report.

Near the south end of Belle Isle Strait, Cartier stopped at a harbor named *Blanc Sablon* by French fishermen for its white sand, and which still has that name today. The fishermen used the harbor for its good beach where they could dry their catch. There he noted to the southwest of *Blanc Sablon* two islands, now named *Isle au Bois* and Greenly Island. Few of the names given to landmark features by Cartier have continued to the present.

Through June of 1534, Cartier continued southwest along the coast naming bays, inlets, and islands he encountered. He reported taking many salmon to add to their larder. They also met a French fishing ship, the only other ship they saw during the entire voyage. Cartier may have felt that he was in undiscovered territory at this time, and perhaps the presence of the fishing ship was a surprise to him. Imagine going to the most remote area in the world and suddenly finding a busload of tourists already there. However, the fishermen posed no threat to Cartier's claims of making the first map and claiming land for the king. After staying close to Cartier's ships for a few days, the fishermen departed and returned to France with their catch.

Cartier's report had little good to say about the land he was coasting, except that it had good harbors. See figure 4.1. The land showed no promise for settlement or agriculture because most places he landed had no soil—just bare rock. When we see how other discoverers of that era tended to exaggerate the potential of their discoveries, we can perceive Cartier as an honest man and most unusual among explorers of the time. He made no effort to aggrandize the meager landscape. The terrain of this landscape had been scraped clean by the Pleistocene glaciers, and the slow development of soil in the cold climate produced barely enough to support a cover of grasses or, in protected places, clusters of stunted spruce trees. Cartier could not in good faith have given a glowing report. The Indians he encountered in this area were of the Beothuk tribe who hunted seals in the area.

At this point, Cartier veered from his southwestward route and sailed eastward to explore the west coast of Newfoundland. It is not clear why he changed his westward direction, but the outcome completed the map of Newfoundland. The west coast had not yet been surveyed, whereas the east coast had been visited by Cabot, Corte Real, Fernandes, and Verrazzano. Cartier soon resumed a southwesterly heading along the west coast of Newfoundland into the Gulf of St Lawrence. Now he was certainly in unexplored waters.

A mid-June storm hurried Cartier's ship along to the Bay of Islands down the west coast of Newfoundland. This beautiful bay forms an excellent harbor lined with two prominent headlands covered by spruce trees. He mapped and named features along the west coast of the island of Newfoundland down to the south end at present day Cape Anguille, which he saw on Saint John's day and therefore named *Cap Sainct Jehan*. From there, Cartier headed southwest across the open water of the Gulf of St. Lawrence. So far, he had seen little of interest along the coasts of Newfoundland—a few good harbors, but little to attract the notice of the king of France. Crossing the gulf he passed the *Îles de la Madeleine* (Magdalen Islands), little more than three

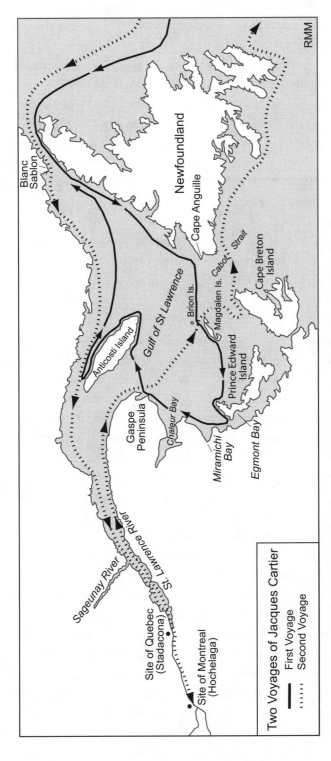

FIGURE 4.1 Jacques Cartier's first two voyages to Canada established France in North America.

rocky bird roosts. The tiny islands, connected by sand spits, were home to thousands of murre and auks. Cartier's crewmen went ashore to resupply and killed more than a thousand birds in less than an hour.

Now Cartier began to describe a much different kind of landscape, and one which showed potential for settlement. The tiny island in the Magdalen group that caught his eye was *Île de Brion*, which is the name Cartier gave to honor one of his supporters. A walk of less than 5 miles would have taken him the full length of the island, and it is only 1 mile at its widest point. It is not clear why he wrote so glowingly of this tiny island, but it was the first piece of appealing land he had seen since leaving France. "Those islands have the best soil that ever we saw, for that one of their fields is more worth than all the new land. We found it all full of goodly trees, meadows, fields full of wild corn and peas bloomed as thick, as rank, and as fair as any can be seen in Brittany, so that they seemed to have been plowed and sown. There was a great store of gooseberries, strawberries, damask roses, parsley, with other very sweet and pleasant herbs." The present view of *Île de Brion* is somewhat less congenial. The trees Cartier saw are now fewer—he may have exaggerated a bit—but the meadows are as beautiful as he said they were. The Magdalen Islands soon became settled by French fishermen. Cartier also told of seeing "sea oxen," his name for walruses. "About the said island are very great beasts as great as oxen, which have two great teeth in their mouths like unto an elephant's teeth, and live also in the sea."

Cartier must have seen the great expanse of open water, now known as Cabot Strait, as he sailed past the south end of Newfoundland, and a short diversion from his actual course would have verified that Newfoundland is not connected to the mainland. This route would have saved considerable distance compared to sailing around the north end of Newfoundland. He could have returned to France through Cabot Strait, but nevertheless, he later retraced his route through the Strait of Belle Isle around the north end of Newfoundland.

Cartier's ships now continued their southwesterly course across the Gulf of St. Lawrence. The first landfall was on the last day of June when he saw the hills near the middle of the north coast of St. Edward Island in the vicinity of New London Bay. Here again he reported a beautiful and fertile land with forests and meadows suitable for habitation—in this instance he was not exaggerating. He wrote of seeing many different kinds of trees, berries, and wild grain. Cartier sailed westward along the coast to the northwest tip of land, North Cape. From there he sailed eastward along the west coast of present day Prince Edward Island toward Egmont Bay, and reported that the coast was lacking good harbors. He turned back to continue westward without discovering that he was traversing an island.

Mapping was a valuable by-product of his effort, but could not take time from the primary objective of finding the prize—passage to Asia. Therefore, there was no need to determine if a land was connected to the mainland, just as it had not been necessary to determine if a strait existed south of Newfoundland. To explore these seemingly vital questions would have diverted him from his instructed objective, and the short sailing season did not allow time for side trips. This same practice continued so long as the primary objective was to find a route through or around the continent. Even the immense efforts of the Royal Navy in the nineteenth century left puzzling gaps in maps that a brief diversion in route might have filled. Only through the search for the Franklin expedition after 1848 did many blank spots finally get mapped. Captain James Cook stands out as a striking exception. His skill in surveying produced a complete coastal map of Newfoundland in 1775 that is still impressive. He produced other detailed maps of coastal areas as well. In that instance, however, a detailed map of the island was the objective.

Sailing west across Northumberland Strait, Cartier followed the mainland coast along what is now New Brunswick. As he went, he mapped Miramichi Bay, which he named *Baye Sainct Lunaire*. This name had the dual purpose of honoring the Breton saint whose day it was (July 2), and also honoring his patron, the abbot of Mont Saint Michel, who had the same name. It has always been good policy to remember those who pay the bills.

Cartier mapped and named *Chaleur Baye*, which remains so named for its appealing warm temperature and attractiveness. He happened to see it on a beautiful sunny day when both the air and water felt comfortably warm. This land pleased Cartier more than any he had seen on this voyage. He described the towering spruce suitable for masts on large ships. The climate was temperate, the soil was rich, and the water was full of salmon. The attraction of this shore soon brought settlers to this congenial place where climate and soils provided ample support for agriculture. Cartier stayed in the area for a week exploring the bay.

While in Chaleur Bay on the south shore of the Gaspé Peninsula, Cartier had his first encounter with Indians of the Micmac tribe. As many as fifty canoes appeared in the bay with men shouting and waving pelts in the air. This gesture indicated they wanted to trade and had apparently already experienced European fishermen's interest in furs. Cartier and his men were out in their boats at the time, and feeling overwhelmed by the size of the party approaching them, beat a retreat to the ship. The faster canoes caught up and surrounded Cartier's boat, and the Indians made signs of friendship and excitement. Cartier's crew felt threatened and tried to wave the Indians away. Cartier's men fired a gun over the Indians' heads and frightened them away. This was not a good start with the locals, but the next encounter one day later went much better. When the Indians again appeared, Cartier sent men ashore with trading items, including knives, other tools, and a red cap for their chief. The crewmen returned to the ship laden with furs.

On July 12, Cartier, reluctant to leave this beautiful place, resumed his search for the passage. After meeting other friendly Indians, these of the Huron tribe, who wanted to trade furs, Cartier raised a cross thirty feet tall on a point of land near the head of Gaspé Bay. He had the words *VIVE LE ROI DE FRANCE* inscribed and attached a plaque showing the *fleurs-de-lys*. Then all crewmen knelt on the ground and raised their arms to heaven. Cartier had thus taken possession of these lands for France. The chief of the Hurons, sensing something significant had occurred to his domain, was not pleased with this event and came aboard the ship to let Cartier know that this was his land and the cross had been erected without his permission. Cartier must have handled this potentially explosive situation with adequate diplomacy, as the chief was soon mollified with presents. The chief came to regard Cartier so highly that he gave him two of his sons to take back to France on the condition they would be returned on the next voyage. The boys were quickly dressed in European clothing, and farewells were made as the expedition resumed.

As Cartier rounded the Gaspé Peninsula, he encountered a fog bank. He sailed north to Anticosti Island, but supposing the island was only an extension of the Gaspé Peninsula, he did not attempt to sail west. After his pleasant experience with Gaspé Bay, Cartier found little to recommend this big island except an abundance of salmon.

Now near the end of July, Cartier sailed almost the entire north coast of Anticosti Island. He held a conference with the senior men of the expedition who voiced concerns about the signs of the changing season. As they had no desire to spend the winter in this region, and sailing farther west would be a risk, they decided the signs should be heeded. On August 2 they began the homeward voyage. They headed again for the Strait of Belle Isle without losing time in further exploration or mapping waters they had already traversed. Their sense of haste may have prevented them from checking on the shorter route through Cabot Strait that Cartier had correctly imagined to exist.

Arriving home in September, Cartier found he had become a man of status. In one voyage of five months, he had mapped almost the entire Gulf of St. Lawrence. He had looked ahead to see that open water continued to the west, giving good reason to return for further exploration. He had unknowingly stood at the mouth of a great river that penetrates deep into North America, opening up the entire Great Lakes region and beyond. The eventual influx of settlement and development was phenomenal. Cartier's glowing report on his exploration provided vivid images of a fertile land waiting for Europeans to arrive. Cartier's task was clearly unfinished, and the time was right for a second voyage to push farther west. Cartier, like John Cabot, had seen the promise of open water to the west and returned home to plan another voyage.

SECOND VOYAGE, 1535–1536

Cartier's success from his first voyage gave him great credibility with the French monarch. The glowing report of his voyage held such favor that the king quickly issued an order granting Cartier power to engage ships and men for another voyage. Cartier was authorized to equip and provision three ships for a period of fifteen months. He worked busily through the winter, and by the middle of May 1535, everything was ready for a new voyage. Again the starting place was St. Malo.

Cartier set sail on the May 19 with three ships, the *Grande Hermine* of 120 tons, the *Petite Hermine* of only 60 tons, and another small ship of 40 tons named the *Emerillon*. In all, a company of 112 men sailed with the expedition. On the *Grande Hermine*, Cartier's flagship, were the two sons of the Huron chief that Cartier had taken home on the first trip. They became guides and interpreters during this second voyage. After only a week at sea, the three ships became separated by storms and then dense fog, and the rest of the crossing was made with each ship traveling alone.

On July 7, the *Grande Hermine* reached Newfoundland and the next day repeated the previous route through the Strait of Belle Isle to *Blanc Sablon* where they anchored to await the other two ships. Almost three weeks later, on July 26, the other two ships arrived safely. There the expedition rested a few days to make repairs and resupply their water and wood.

Once underway, they traversed a route through the Strait of Belle Isle by the same route they had taken the previous year. Cartier took pains to record his positions as well as the water depths and bottom conditions. He again described landmarks for the benefit of future mariners. He noted that dense forests extended down to the shore, but overall the terrain seemed uninviting to him. He headed for Anticosti Island—his most westerly point the previous year—and from there he embarked into new territory up the St. Lawrence River. As they traveled along the north coast of the Gaspé Peninsula, their Indian guides told them they were entering the kingdom of Saguenay along a great river that leads to the interior of Canada.

The discovery that they faced a river of fresh water and not a sea route to China disappointed Cartier. He knew that any discovery of an inland kingdom would seem pale relative to finding a route to Asia. Cartier sailed along the western coast of the Gulf of St. Lawrence just to be certain he had not overlooked a passage.

Finally, on August 15, Cartier could delay no longer and entered the mouth of the St. Lawrence river. The sailing season was nearing an end, and Cartier had to decide whether to return to France with little to report or to stay for the winter. The importance of a successful voyage, i.e., finding something of great interest, prompted him to stay. It would be far better to stay and produce an extensive report than to return with nothing. Perhaps he also hoped that the great river would in fact become an important passage to the Pacific. By September 1, he passed the mouth of the Sague-nay River. Here Cartier reported seeing a beautiful river about a mile wide at the mouth, but wider upstream between high, steep cliffs. The shallow soil supported a dense growth of large pines.

As Cartier continued up the difficult tidal reaches of the St. Lawrence, he named the islands and recorded that they appeared rich, fertile, and covered with trees. One of these he named *Isle-aux-Coudres* for the many filbert trees he saw there. This island still carries the name given by Cartier. The Indian guides told him that this area marked the beginning of the country they called Canada. The guides may have used the term Canada to mean a settlement, but Cartier understood it to mean the entire region.

At the site of present day Quebec lived hundreds of Huron-Iroquois Indians in a settlement called Stadacona. This beautiful land supported forests of maple, birch, ash, and elm along the banks of the river. In clearings the Indians grew corn and peas. In all respects, Stadacona appeared to be a thriving and prosperous place, but strangely, just sixty-four years later, Samuel de Champlain visited this area and found the spot deserted.

The two Indian guides provided a vital connection with the Huron-Iroquois in-habitants of the area. Not only did they know the language, but as sons of their chief they also were known by most of the Indians, and everyone knew that they had been taken by the Europeans. Once the word spread that the white men had returned with the two sons, canoes filled with men and women converged on the ship. The next day, the chief, Donnacona, came alongside the French ships and made a lengthy speech. When the sons told the chief of the wonderful things they had seen in France, the chief put his arms around Cartier's neck to express his welcome. This grand display set the tone for the visit. Cartier was concerned that the two Indian guides had rejoined their people but now kept their distance from him. Cartier had to persuade them to continue with him up the river to Hochelaga. Cartier also noticed that the chief began staying at a distance as well, and Cartier developed a feeling of uneasy caution. Probably the two guides wanted to be sure they did not find themselves returning to France against their will.

Cartier decided to confront the chief about his reservations and approached him with several armed sailors. The chief, speaking through the interpreters, criticized Cartier for approaching with arms while they were unarmed. Cartier tried to explain that it was customary to carry arms in Europe. Nevertheless, Cartier and Donna-cona parted in good spirits and with assurances of continued friendship. Cartier realized that the two guides had created the issue and not the chief. The chief returned the next day with about five hundred Indians and an entourage of impor-tant leaders of the tribe. Cartier invited the leaders aboard his ship and provided a feast with presents for the guests.

Donnacona made a great effort to persuade Cartier not to continue up the river to Hochelaga. He proclaimed that there would be nothing of value there and Cartier would like it better in Stadacona. Cartier argued that his king had ordered him to push as far as possible up the river. Donnacona's next move was to invite Cartier ashore where he offered Cartier a little girl, his niece, and two young boys as a gift if he would stay away from Hochelaga. Cartier was not interested in taking the children, but he was convinced to accept the chief's gesture of goodwill. Cartier responded by presenting some valuable gifts to the chief, including two swords and a brass basin and pitcher. Donnacona's last request was to hear the ships' cannons; Cartier complied by firing twelve cannons and creating a great stir of excitement among the Indians gathered.

The Indians of Stadacona made one more attempt to dissuade Cartier from going to Hochelaga. They put on a big display of a devil-like figure appearing in a canoe with a message that Cartier and his men would all die if they went to Hochelaga. This ruse failed to convince Cartier, and the Indians finally resigned themselves that the Frenchmen were leaving. Probably the Indians felt the Frenchmen were heaven sent, and if they went to Hochelaga they would never return to Stadacona. In order to navigate in the narrowing river, Cartier left the two larger ships moored near Stadacona and took the smallest, the *Emerillon*, with a company of fifty men to Hochelaga.

Nine days farther up the St. Lawrence, Cartier's ship came to a place where the river broadens into a lake now called *Lac Pierre*. Cartier described the beautiful maple and oak forests in glorious autumn color. He told of seeing countless geese flying and birds of many kinds in the forests. He thought a lovelier country could not be found. Indians along the way came to his ship in canoes bringing fish and food in exchange for presents from Cartier. His reputation for friendliness and generosity apparently preceded him. An Indian chief along the way brought two children as gifts for Cartier, one a girl of 8 and the other but a toddler of 2 or 3. Cartier accepted the 8-year-old as a token of the chief's goodwill and took her with him to Hochelaga and later back to Stadacona.

Upstream from *Lac Pierre*, the river narrowed and had many sand bars. As the Indians told him that Hochelaga was still three days upstream, Cartier decided to leave his ship and proceed in two boats. He took twenty men, and, leaving the others to watch the ship, he continued toward Hochelaga. On October 2, they arrived at the site of the present day city of Montreal, then the head of navigation on the St. Lawrence. Cartier wrote that a thousand Indians came to the shore, leaping and singing in a welcoming greeting with offerings of fish and cornbread. Many held up their children wanting Cartier to touch them. If this display did not make Cartier feel like a god, it certainly must have impressed him with the potential for France in this country. Cartier set a good example with the two cultures having a harmonious first meeting.

To the Indians of Canada, the French sailors with their light skin, strange clothes, and beards—arriving in ships with immense sheets of sail—presented a most impressive sight. On the other hand, the French, surrounded by a multitude of exuberant, dancing, brown-skinned people wearing feathers and animal skin clothing, must also have thought this the most exotic sight of their lives. Such a moment could never be repeated for either culture as their futures were forever changed.

The site of Hochelaga stood amid many agricultural fields of corn, millet, and peas. It was near a mountain which Cartier named *Mont Royal*, still a prominent landmark in Montreal. The settlement of Hochelaga was surrounded by three courses of timber wall with one entry gate. Cartier counted about fifty wooden houses, each about fifty paces long and twelve to fifteen paces wide. The houses were placed around a large central square, and each house had many rooms around a central court where the Indians built a fire.

Cartier wrote that the chief had shrunken limbs from an illness Cartier described as palsy. He was wearing a wreath of porcupine quills and was carried into the central square and placed on a stag's skin. The chief showed his arms and legs to Cartier and wanted him to touch them, which Cartier did. The chief then transferred his wreath to Cartier's head as a symbolic gesture of gratitude. Then the Indians brought a number of other diseased, crippled, and blind men, asking Cartier to touch them as if he had godly powers to heal. Rising to the occasion, Cartier read from the Gospel of John, and as he touched each man he prayed to God that these people might receive Christ. The Indians sat silently and imitated Cartier's gestures. Cartier then gave gifts to all, including women and children.

Some of the Indians took Cartier to the top of *Mont Royal* to express to him the vastness of the area which could be traveled for months on water. Doubtless they referred to the Great Lakes. Also, they pointed to a silver chain that Cartier wore and a dagger handle made of copper that appeared as bright as gold, and indicated that such metals came from the land beyond the river. Cartier understood this to mean that deposits of silver and gold were known to the Indians.

Cartier left Hochelaga and returned down river to Stadacona to reunite all of his expedition again after a separation of twenty-two days and to prepare for winter. Cartier may have been the first European explorer since the Norsemen to winter in the cold regions of North America. They built log shelters inside a wooden stockaded fort with a moat and drawbridge to protect themselves against possible attack from the Indians at Stadacona. The French kept fifty men on guard in shifts through the nights. One item of note that Cartier commented on was his introduction to tobacco, which the Indians dried, powdered, and smoked in a pipe. Cartier thought it a most unpleasant taste.

A disastrous scourge of scurvy hit both Indian and French populations in the winter of 1535–1536.[4] By December, eight of Cartier's men had died and fifty lay at the point of death. Only three or four men had not yet been struck by the disease, although Cartier himself remained in good health. One day, one of the interpreters from Stadacona approached him. Cartier knew that the man had been ill with scurvy, but now he was obviously in good health. On inquiry from Cartier, the interpreter told him of a drink made from the leaves and bark of a tree. Cartier, saying only that one of his men was ill, asked the interpreter to bring him the branches of the tree. Later a woman came with some branches with instruction to boil them and drink the liquid twice a day. The recovery of the French sailors was so rapid that Cartier declared it to be a miracle. He said all the doctors of France could not have done as well. All the sailors were restored to health. The tree in question was probably the Eastern Arborvitae, a tree rich in vitamin C and a known antiscorbutic.

In the spring, a large number of Indians from other areas arrived and stayed near Stadacona. Cartier feared this would not bode well for the safety of his expedition, and when the ice in the river began to clear, he made preparations to leave. Cartier hatched the idea to take Donnacona back to France. The chief had told him so many tales of odd races of humans that lived beyond their country—men without stomachs and men with only one leg—Cartier thought the chief should relay these incredible stories to King Francis I in person. Unfortunately, he failed to mention his idea to Donnacona.

Cartier erected a cross thirty-five feet high engraved with the coat of arms of France and the inscription, *FRANCISCUS PRIMUS DEI GRATIA FRAN-CORUM REX REGNAT* (Francis I, by the grace of God King of the French, is sovereign). At this moment, France and French culture became a permanent feature of the Canadian landscape. Then, in celebration of the event, Cartier invited Donnacona and a few others on board his ship and immediately made them prisoners. This created a great uproar, both among the captive Indians and the people standing on shore. Cartier managed to calm the situation and convinced Donnacona to stay by use of his seemingly endless supply of gifts and a promise to return him the next year. Given the good relations up to that time, Donnacona might have agreed to go to France without coercion.

Finally on May 21, 1536, the expedition began the voyage home. Because a number of sailors had died from scurvy, Cartier abandoned the *Petite Hermine*, and with the combined crews manning the *Grande Hermine* and *Emerillon*, he returned to St. Malo.

THIRD VOYAGE, 1541–1542

Cartier's report of his second voyage was also well received. Francis I listened attentively as Cartier related his adventures and the sights he had seen. Chief Donnacona had his time before the king and told of wonderful things about his country. The king ordered that Donnacona and the other captives should be received into the faith, and records show that three Indians brought by Cartier from Canada were baptized on March 25, 1538, three years after leaving their homeland.

Cartier proposed a third voyage, and although the king was interested, other pressing matters of state—conflict with Spain—diverted his attention and resources.[5] In 1538, a truce was settled with King Carlos V of Spain, and Francis I again looked to his new discovery in Canada. Not until 1541 did Cartier again set sail, but now the conditions and the objective had changed radically. This voyage was not intended to discover a passage to China or to map new lands, but to establish a settlement and find natural resources. Cartier was to command the ship, but one of the king's favorites, Jean François de la Roque de Roberval, was to be governor of the settlement. Roberval had the task of acquiring provisions, including plenty of cannon and powder, and of recruiting sailors and colonists for the venture. He found few people interested in becoming colonists and arranged to take convicts as potential settlers. Soon a full roster of recruits, chained and under guard, were brought to the port at St. Malo.

On May 23, 1541, Cartier set sail with five ships. Roberval was not fully ready at that time and planned to leave later. The five vessels arrived at Stadacona on August 23 and found the Indians eagerly awaiting their long missing chief Donnacona. Unfortunately Donnacona and all his fellow captives except a little girl had died in France. This created a very difficult moment, which Cartier handled by telling the gathering that Donnacona had indeed died, but all the other captives had become wealthy lords in France, and chose not to return. The new chief of Stadacona was apparently pleased with this news, as his own position would not now be in question.

Cartier's convicts built forts, planted gardens, and established a settlement nearby named *Charlesbourg Royal*. Also, the men loaded bags of useless iron pyrite and quartz crystals onto the ships thinking they had found gold and diamonds. After a winter at the settlement, relations between the Indians and the colonists grew hostile. The Indians did not make an attack, but Cartier decided his part of the agreement as an advance party was ended, and he set sail for home in June of 1542, leaving a precarious colony behind. In the vicinity of Newfoundland, he met Roberval, who was just now arriving with his provisions and artillery. Cartier refused to return to *Charlesbourg Royal* with Roberval and proceeded to St. Malo. The next year, the struggling settlement was abandoned because of illness and the hostility of the Indians toward the newcomers, and the survivors returned to France.

No other attempt at settlement by Europeans was made in the area until 1608 when Samuel Champlain established Q uebec. Cartier never made another voyage and spent the rest of his days on his estate near St. Malo until his death in 1557 at the age of 65.

PART II

England Reenters the Game, 1576–1632

A wet sheet and a flowing sea,
A wind that follows fast
and fills the white and rustling sail
and bends the gallant mast.
　—ALLAN CUNNINGHAM

He that would go to sea for pleasure would go to hell for a pastime.
　—PROVERB

5

Ships, Navigation, and Mapping in the Sixteenth Century

LIFE ON THE SHIP

Any boy who intended to be a sailor first became an apprentice no later than age 14.[1] The boy's parents or guardian, if they had the means, paid a ship's master or first mate ten or twenty pounds to train the boy for up to nine years as an unpaid apprentice.[2] In return, the ship's master would see that the boy learned everything about sailing. Boys for whom a payment was made could expect rapid advancement after the apprenticeship ended, and they could anticipate becoming a first mate or a ship's master by their mid-20s.

Boys from poor families could be accepted without a payment, but they would have little hope of advancing above the level of able seaman after the apprenticeship. Often they became little more than the master's servant without much training beyond menial duties. Whether wealthy or poor, apprentices were bound to years of unpaid service to the ship's master. After the apprenticeship, they began to receive some wages, which were paid in a lump at the end of a voyage.

In 1733, the Royal Naval Academy began training officers. A boy could enter at age 13 at the expense of parents or a patron and spend four years training before beginning his experience at sea. The Academy provided an alternative to the traditional method of training, but sea captains tended to have a higher regard for men trained wholly at sea, rather than at the Academy. Today, the Academy, now called the Britannia Royal Naval College, is the primary training for officers, but regular seamen may still advance into officer ranks.

A sailor's work included several types of tasks. First, he had to assist in the preparations for a voyage. Most ships needed some repair after a previous voyage. Damage from storms to masts, sails, and hull required replacement, sewing up sails, or scraping, caulking, and tarring hulls and decks.

As the time neared for departure, sailors would be loading and stowing the food, water, and other ship's stores. Food consisted of salted beef and pork, biscuits, and dried peas. Often stalls and coops for pigs, chickens, sheep, or cattle needed to be built. Other stores included all the necessary supplies: candles, firewood, brooms, buckets, rope, pots and pans, tools, beer, wine, and dozens of items needed for self-sufficiency during the voyage.

The food loaded for Martin Frobisher's second voyage in 1577 was considered sufficient for 120 men for up to four months. Included in the food list on Frobisher's ship were: one pound of biscuit (also known as hardtack, tack being a sailor's slang for food) per man per day; one gallon of beer per man per day; one pound of salt beef or pork per man on meat days, plus one dried codfish for every four men on fast days; oatmeal and rice were loaded as backup in case the fish supply ran out; one quarter pound of butter and one half pound of cheese per man per day; honey (sugar was still a rare luxury then); a hogshead (a 64-gallon barrel) of cooking oil; and a pipe (equal to two hogsheads) of vinegar. These sailors ate and drank well, as they must with the amount of energy expended in their daily work. Upon reaching land, they planned to hunt game animals and birds to supplement the meat supply. Also, by carrying fishing gear, they caught cod and salmon when sailing in good fishing areas. With the danger of loss from spoilage of both food and beer, and leakage from barrels, they had to keep a close watch on their stores to be sure the supplies lasted. The salted meat needed to be dragged by rope through the sea water for several hours before cooking to freshen it a bit. If fresh meat was supplied for the voyage, it had to be eaten in the first few days. Livestock on board could make fresh meat later in the voyage as well, but many ships had no room for animals.

The hardtack biscuit was made of flour and water, and its main advantage was that it had a very long shelf life. The hard, dry biscuit had to be moistened with water or beer to make it easier to chew. Usually, it already had weevils in it before it was loaded onto the ship because it was made months in advance and stored for provisioning ships. A typical menu for seamen might be salted meat with peas porridge consisting of dried fish in a thick mixture of pea soup, accompanied, of course, by a weevily biscuit. Although this sounds like a wholesome meal, as the weeks went by, the meat might have spoiled, the butter turned rancid, the beer turned sour, and the biscuits reduced to dust by the weevils. Officers were usually provided with wine or spirits rather than beer, and some ships supplied the sailors with a regular ration of rum as well. A cook box was set up on a bed of sand or rocks below the forward deck, and sailors would go there to get their bowls filled and then eat wherever they could find a spot to sit. Some larger ships eventually began to provide a mess area where the crew could sit and eat.

The sailors had new tasks when the ship weighed anchor to get underway. This stage of ship life required the efforts of the entire crew until the anchor was secured, sails set as directed by the ship's master, and all gear and cargo stowed. A watch schedule was set in the first evening, allowing part of the crew to sleep while others tended the ship. Watches were four hours long starting at eight in the evening, though all hands were expected to busy themselves during the daytime. In an emergency, all hands would be summoned, even in the night. It was a demanding schedule with only brief times during the day for relaxation. At each watch, one man was assigned at the helm, and one to a crow's nest on the mast from which to keep lookout for land, ice, floating objects, or other ships.

While underway, a sailor's work comprised some real hazards, especially climbing the shroud lines up to the yards and standing on foot-ropes as they attached sails to the yards. Sails could be furled (rolled up and secured to the yard) or unfurled, reefed (shortened to a length appropriate to the strength of the wind) or unreefed. A fall from the yard was almost always fatal whether the sailor fell into the sea or onto the deck. If he fell into the sea, he usually drowned before the ship could rescue him. Normally there was little if any effort made to rescue, especially if the water was very cold, as he would live only minutes. Often the hazardous work was done at night in the midst of a howling storm, when the sails needed to be reefed up to compensate for the strong wind. This meant that sailors had to know the rigging well enough to work in total darkness with fingers numb from the cold.

Work on the deck during a storm was almost as dangerous due to the chance of being washed overboard as the sea spilled onto the deck. Everything not tied down was in danger of washing away. Every change of wind or change of course required resetting some or all of the sails. After the storm, there were usually repairs needed, such as replacing a broken yard, or patching or replacing a damaged sail. Damage to the hull could cause a leak, requiring hand pumping and repairing damage. Ships tended to leak somewhat, even though caulked and tarred. When nothing urgent was happening, there was always scrubbing the deck or drying out water-soaked goods and clothing. There was a constant need for scraping all of the wood surfaces and masts and retarring them.

A sailor usually provided his own clothing consisting of a wool pullover shirt with hood to protect from wind, woolen trousers, and long woolen stockings. The color of most clothing was white, red, or blue. They had shoes, but often went barefoot to

avoid slipping on decks and ropes. Officers usually wore jackets and hats that distinguished them from the sailors. No clothing provision was made for bad weather
unless the sailor brought it himself. Some sailors had up to six changes of clothing to
allow for drying of soaked clothes and to avoid sleeping in wet clothing.

Hammocks for sleeping were discovered by Columbus among the natives in the
West Indies, but they were not widely used on ships until almost one hundred years
later.[3] During the sixteenth century, sailors slept wherever they could find a vacant
place on decks or cargo. Bunks and cabins were provided for officers. Living space on
a ship was small and crowded. For example, the main deck of Columbus's caravel,
Niña, was 66 feet long and only 17 feet wide. Sailors often slept on the deck in the
bow, or below in bad weather. In heavy seas, the hatch had to be secured to keep the
water out, and the quarters below deck became dark and airless. The smell of
unwashed men, animals, and spoiled food made the need for airing out in fair
weather very important. In small ships, the height between decks was usually less
than five feet. Although men's average height in the sixteenth century was somewhat
less than today, few could stand erect below deck.

Everyday dangers and reports of ships lost at sea without trace were real and frightening to men traveling the oceans. Such stories gave rise to rumors about the supposed causes of the loss and the suspicion that misbehavior of the crew or of a crew
member was the real reason a ship was swallowed up by the sea. The practice of religion while at sea became a way to provide protection against those dangers. Sailors
felt they were in the hands of the Almighty and made frequent prayers to God and
to the Virgin Mary for protection. Some ship's captains forbade swearing, blas-
pheming God, filthy stories, or any ungodly talk. Also, gambling with dice or cards
and fighting were forbidden on some ships. Morning and evening prayer services,
with readings from the Bible, were often part of the required daily routine. Injunc-
tions against foul language on board ships lasted through the nineteenth century.

Superstitions, sometimes mixed with religion, also played a role. A ship should
not sail on Friday because it was the day Christ was crucified. One must never lay a
hatch cover upside down—no reason is known for this. When building a ship, a silver coin must be placed under the base of the main mast. One must never destroy a
printed page as it might belong to the Bible. Many sailors were illiterate and could
not be sure if a page might be from the Bible, so they took the safe position that all
printed pages were scripture. Some shipbuilders carried a burning fire brand through
every part of a new vessel to drive out evil spirits.[4]

A sailor's work was extremely hazardous and injuries were common, but the greatest danger aboard ships on long voyages in the sixteenth century was scurvy.
Unknown to mariners at the time, the cause of scurvy was a diet containing
insufficient vitamin C. Any fresh fruits, vegetables, or meat on board were soon

consumed, and the supplies during the rest of the voyage were dangerously deficient in this important vitamin. After about six weeks of salted meat and hardtack, the first symptoms began to appear—swelling of the gums and loosening of teeth, then blotches on the skin, followed by a deep lethargy leading to death. Consumption of vitamin C could quickly correct these symptoms—except for the death part.

Although scurvy was known among the earliest mariners and fruits were thought to relieve the illness, it was not until 1747 that a systematic experiment was conducted by James Lind, a British naval physician, to study the effects of citrus fruits on the disease. Lind, like others of his time, believed that scurvy was caused by putrefaction in the body, and that the acidity of citrus fruit corrected it. Even though Lind demonstrated the benefits of citrus, it was not until 1795 that lemon juice was required to be carried on ships of the English navy. In the mid-nineteenth century, lemon juice was replaced by lime juice because of lower costs. However, they discovered that limes have only half the antiscorbutic value of lemons.[5]

Despite the poor pay and all the hazards, sailors felt bonded to life at sea. Eric Newby told of both the dangers and the thrill of sailing when he wrote of his experience as a young crewman aboard a big square rigged ship in 1938. Sometimes fearing for his life, he wrote, "At this height, 130 feet up, in a wind blowing 70 miles an hour, the noise was an unearthly scream. . . . [T]he high whistle of the wind through the halyards sheaf, and above all the pale blue illimitable sky, cold and serene, made me deeply afraid and conscious of my insignificance." However, standing on deck on a fair day, Newby described the joy of sailing. "As time passed, the ship possessed us completely. Our lives were given over to it. A hundred times a day each one of us looked aloft at the towering pyramids of canvas, the beautiful deep curves of the leeches of the sails and the straining sheets of the great courses, listened to the deep hum of the wind up the height of the rigging, the thud and judder of the steering gear as the ship surged along, heard the helmsman striking the bells, signaling a change of watch or a mealtime, establishing a routine so strong that the outside world seemed unreal."[6]

Among ships sailing the oceans in the fifteenth and sixteenth centuries, the caravel and the carrack were the most often used. During the seventeenth century, the galleon superseded the carrack. The caravel was used by Columbus, *Niña* and *Pinta*, and by John Cabot, *Mathew*. It was typically a three-masted (occasionally four-masted) ship with square rigged sails on the front two masts and a triangular lateen sail on the mizzen mast (figure 5.1). In its early developmental stages, the caravel was a fishing boat with all sails rigged as lateens. The caravel was a small, usually less than one hundred tuns,[7] fast, and easily maneuvered ship that could head closer into the direction of the wind than most ships of the day. Its length averaged about 80 feet, though some reached 100 feet.

FIGURE 5.1 The caravel began with lateen sails and evolved into a three masted vessel rigged with square sails on the forward masts and lateen (triangular) on the mizzen (rear) mast. This ship, preferred by many mariners in the 15th and 16th centuries, averaged seventy to eighty feet in length, but some were as long as 100 feet. Wikimedia Commons.

FIGURE 5.2 The carrack was larger than the caravel, but with sails rigged in the same way. It had high castle structures both fore and and aft. The carrack depicted was Magellan's *Victoria* as displayed on a 1589 map of the Pacific Ocean by Ortelius. Wikimedia Commons.

A carrack, such as Columbus's *Santa Maria*, was somewhat larger than the caravel and saw common use in northern and southern Europe through the fourteenth to seventeenth centuries. The three-masted carrack rigged sails were similar to the caravel, but had high castles for living and storage space both fore and aft. Although the carrack lacked the speed of the caravel, it could carry more crew and cargo. The largest carracks, shown in figure 5.2, were known to be about 1,200 tuns. This type of ship was the mainstay of commerce and exploration until the development of the galleon in the seventeenth century.

NAVIGATING THE SIXTEENTH-CENTURY SHIPS

European sailors developed skills for navigation in the Mediterranean Sea when Greek and Phoenician sailors began to observe the stars and note their risings, crossings of the zenith, and descents in the west. Also they noticed that stars in the most northern quarter merely rotated like a wheel without setting below the horizon, and that the rotating array of stars defined a pivotal point, which, prior to the tenth century, had no actual star at its center.[8] The seven stars of greatest interest because of their high visibility and their nearness to the pivot point were called, variously, the Bear, the Plow, or the Wain (wagon). Today we call it the Big Dipper or *Ursa Major*, and it is the first star group that most children learn to identify at an early age. Some maps of the fifteenth and sixteenth century labeled the northern regions of the world by the Latin term *Septentrionalis*, referring to the region under the seven stars of the Big Dipper, or the seven oxen of the Plow. The Little Dipper, *Ursa Minor*, being closer to the point of the turning axis, later became more important to navigators.

The pivot point in the sky around which the stars appear to turn is not a fixed point. Through the centuries, it very slowly migrates along an elliptical path caused by the slight wobble of Earth on its axis. Through time, the axis of Earth's rotation points to different locations in space. Astronomers have been able to retrace this change to show how the stars appeared at different times in the past. In 1000 BC, the star Polaris was about 15 degrees away from the pole and was of no particular interest to star watchers or early navigators. At that time, the star of the Little Dipper closest to the axis of rotation was at the outer edge of the cup of the Dipper, the star Kochab. See figure 5.3.

Mediterranean mariners learned to find their way by knowing the winds, learning to read the sea currents and waves, measuring depth, sampling the bottom to help identify location, and reading the stars. They had no instruments for measuring anything except a weighted line for measuring depth—running aground has always been a source of anxiety for sailors. Contrary to general belief, sailors—even without navigational instruments and charts—could find their destinations without hugging the shorelines. Actually, the fear of being drawn into the shore by currents and waves prompted them to stay well away from the shore while under sail. They were at ease sailing anywhere within the Mediterranean Sea, making cautious ventures into the Atlantic to reach other ports in Europe and eventually westward to the Azores. They knew the stars by night, observed the sun by day, and were guided by

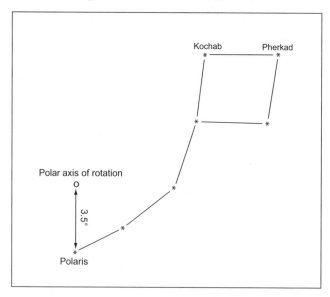

FIGURE 5.3 This diagram shows the offset of the star Polaris from the axis of rotation in approximately the year 1600. Navigators measuring angles for their computations of latitude learned the appropriate corrections based on the various positions of the "guard stars" *Kochab* and *Pherkad*. In the position shown *Kochab* and *Pherkad* form a horizontal line above Polaris and the correction to angles read on the astrolabe or quadrat would be +3.5 degrees. Visualize the seven stars of *Ursa Minor* moving counterclockwise around the point of rotation. After 180 degrees of rotation, Polaris would be above the axis of rotation and the mariner would make a correction of -3.5 degrees. Drawn by R. M. McCoy.

known landmarks as they approached land. The one tool they often used was a weighted lead for measuring depth and for bringing up samples of the sea bottom. From the bottom sediments mariners might tell the nearness of land, learn the presence of a nearby river mouth, and know when they were nearing a particular port. Use of the weighted line for depth measurement and sediment sampling near coasts continued into the twentieth century.

Around the year AD 1000, significant changes in European culture took place. Most important was a revival of interest in mathematics and astronomy after many years during which the knowledge of the Greeks, Persians, and Arabs was lost or hidden. As a few men mastered the skills of astronomy and geometry, they began to track and predict the movements of the stars, and they were regarded by many as magicians dangerously involved with the Dark Arts. To aid in their observations, a simple angle measuring instrument, the astrolabe, was developed. See figure 5.4. This instrument had an obvious application to navigation for determining latitude, but mariners were slow to adopt it to their needs. There was little communication between astronomers, who didn't sail, and mariners, who were not scholars nor even literate. Mariners in general were satisfied with their ability to sail to a destination

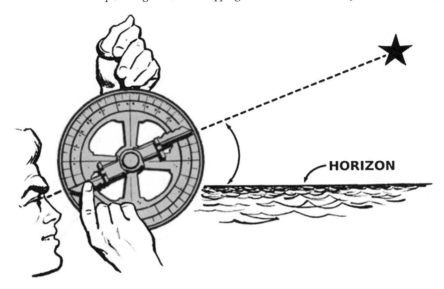

FIGURE 5.4 The astrolabe was used by astrologers and astronomers as early as the seventh century. Seamen adopted the astrolabe in the fifteenth century when they began to venture farther into the oceans. A scale ranging from 0 degrees to 90 degrees was engraved on the disk. Wikimedia Commons.

and were not inclined to change, so about two hundred years passed before the astrolabe came into general use at sea.

Soon after AD 1100, the unusual properties of a quite common iron oxide mineral, magnetite, became known to Mediterranean sailors. Curious observers saw that magnetite would align itself with north. This was a mysterious phenomenon because nothing was yet known of Earth's magnetic field. They also found that rubbing an iron needle against the lodestone passed the mysterious property to the needle. By sticking the needle through a piece of straw it would float in a bowl of water; then the needle would orient itself toward the north. The fact that the needle would point approximately north could not be explained except by some force in the polar constellations acting on the lodestone or on the needle that had touched it. By this time, Polaris was within six degrees of the stellar rotation axis and appeared to be motionless while other stars circled around. This made Polaris of much greater potential importance to navigators. However, as with the astrolabe, mariners did not adopt the magnetic needle quickly, but gradually began to use it only when visual methods failed, such as during cloudy weather.

In order to protect the needle from wind and make it visible at night, the floating needle was put into a protective housing with a lamp, and thus the binnacle was created and is still in use today.

In the thirteenth century, Venice and Genoa were the sailing centers of the Mediterranean Sea. They developed more sophisticated navigation methods using techniques that were well beyond the abilities of the average illiterate seaman, requiring that pilots and shipmasters have an education beyond knowledge of the magnetic needle. Italy became the source of new navigational methods throughout the Mediterranean. A primary development was the expansion of the compass rose (originally called a wind rose) from eight to sixty-four points. This provided the navigator greater precision in describing directions. Rather than describing a course as a little north of east, the navigator could use the more specific term northeast by east, and could align his ship along a specific course halfway between northeast and east on the compass rose.

For this greater precision, the compass had to be something more than a needle floating in a bowl of water. Instrument makers by the end of the twelfth century had balanced the needle on a metal pivot point overlying the compass rose enclosed in the binnacle. As this device made its way north and into the lives of English, French, Dutch, and German sailors in the late twelfth century, the compass rose was reduced to thirty-two points, and the name "compass" came into common use from the Italian *compasso*. See figure 5.5.

As the coastlines became well mapped, a navigator could draw a line from his origin to his destination on the portolan chart, a name derived from the Italian *porto*, meaning port. He could then look for a parallel compass line among the radiating lines on the chart and use that bearing for his voyage (figure 5.6). If none of the thirty-two radiating compass lines on the chart exactly matched his intended line of sail, he used the closest one he could find. Now the mariner had the essential tools for a voyage—compass, chart, depth lead, and a sand glass for keeping time. Keeping track of time was essential for estimating how far the ship had sailed during each watch.

A compass needle points directly to true north in only a few places on the earth. In the rest of the world, navigators must know how much the compass deviates—the magnetic declination—from true north. Variations in the earth's magnetic declination could be ignored by early navigators in the Mediterranean without problem as the area has too little variation between magnetic north and true north to create a serious navigational problem. Sailing north along the coast of Europe or south along the west coast of Africa also caused no need to consider magnetic variation, because the variation in magnetic declination is small in those regions as well. Later, as ships began to sail westward across the Atlantic Ocean, mariners noticed troublesome differences between direction to the Pole Star and the direction their compass needle pointed. Over time, many ships reported their observed deviation, and eventually navigators learned to compensate their compasses as they sailed long distances.

The next major development for navigation was improved instruments for measuring angles of stars above the horizon. This soon led to the development of

FIGURE 5.5 The compass rose was originally called a wind rose and directions were spoken of as "winds." The circle is divided into thirty-two points of 11.25 degrees each. The compass rose would be placed under the compass needle in its box, or binnacle. Wikimedia Commons.

the cross-staff, which consisted of a wooden staff with a sliding cross piece set perpendicular to it (figure 5.7). The sliding piece would measure the height to a star by sliding it until the top of the cross piece was on the line of sight to the star while the opposite end stayed fixed on the horizon. A description of the cross-staff first appeared in 1342, but must have been in use for some time earlier. Because most mariners were not yet acquainted with the idea of "degrees" or even of "latitude," the earliest cross-staff was not graduated in degrees, but was marked with known locations. A navigator would move the cross piece along the staff to a mark for the desired destination, then sail north or south along a meridian until the cross bar just fit between the horizon and the Pole Star (or the sun, as the case may be). He would then be at the latitude of his intended location and could then sail along that course to his destination. Even after degrees and latitude became familiar, this method of navigation, called "running down

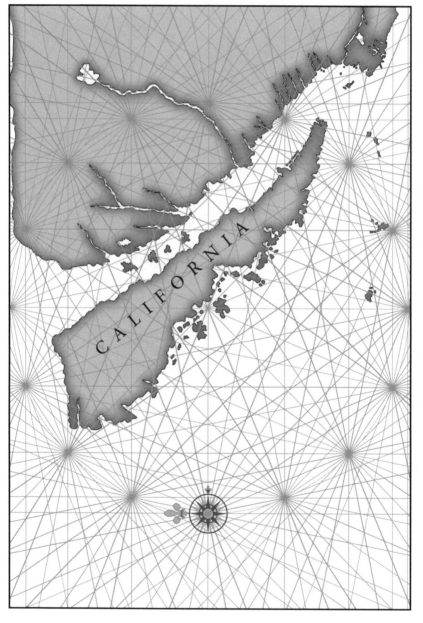

FIGURE 5.6 The portolan chart consisted of compass roses with extended lines. A mariner could place a straight edge between his port and his destination, then find a line on the chart that paralleled his route. By observing the compass direction of the line he could set his course. This map, made by Dutch cartographer Johannes Vingboons in 1650, shows California as an island. Wikimedia Commons.

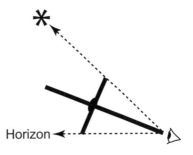

FIGURE 5.7 The cross-staff, or Jacob's staff, was used for measuring angles to the sun or a star. The cross piece was moved until its lower end aligned with the horizon and the top aligned with the star to be measured. The staff was graduated in degrees to correspond with the vertical angle to an object being sighted. For stars at a low angle above the horizon, a shorter cross piece would be used. Drawn by R. M. McCoy.

the latitude," was in practice for many Atlantic crossings in the sixteenth and seventeenth centuries.

Navigators of this time knew that the Pole Star was several degrees off the polar axis of rotation, and those interested in obtaining the best possible measure of latitude soon learned a rule for estimating a correction to their angle measurements. They noted the position of two stars (Kochab and Pherkad, called the "Guards") that form part of the Lesser Bear or Little Dipper. This constellation makes a complete counterclockwise circuit around the polar axis each day. A mariner had to know how to use the position of the Guards to determine how many degrees the Pole Star was off the axis point, i.e., true north. For example, when the Guards appeared horizontal, Polaris was about three and one-half degrees below the actual polar axis, and the navigator added that amount to his astrolabe reading to calculate his latitude. Refer to figure 5.3. By knowing the angular distance of the Pole Star from the axis point in the sky for each position of the Little Dipper, a mariner could make corrections for measurements of latitude. As a memory aid, the navigator could visualize a human figure with its torso at the center of rotation and note the position of Kochab relative to the head, shoulders, hands, and feet of the human figure. During the day, the navigator would measure the angle of the noon sun above the horizon to compute latitude. This computation required knowing the latitude where the sun was directly overhead on any given day of the year. In the fifteenth century, this information was widely available to mariners.

By the fifteenth century, the astrolabe, shown in figure 5.4, had become a tool used not only by astronomers, but also by mariners. This device was a circular piece of metal with a thumb ring placed so the instrument would hang vertically. The disk was graduated in degrees with 0° at the horizon and 90° at the vertical position. On one side was a sighting bar, with pinholes at each end, that swiveled around the center. The observer held the instrument by the thumb ring, sighted a star or the sun through the pin holes, and read the angle. Simple though it sounds, sighting a tiny point of light through the pinholes on a moving ship was not so easy. For more accurate measurements, the navigator took every opportunity to make observations on land. When available, these land-based measurements helped establish more reliable location points for making maps.

There were several approaches to coastal mapping. The most common method in the fifteenth century was no more than keeping a log of the ship's position along with features visible on shore at frequent intervals. The logbook then became the source for the cartographer's map. A more sophisticated approach to coastal surveying was a running traverse off shore, taking compass bearings to objects on shore. As the ship moved along, an assistant would observe and sketch the outline of the coast, putting in inlets, islands, and harbors. Also, occasional depth soundings would be taken along the traverse. No doubt some relied more heavily on sketching by observation, taking only occasional ship positions. A traverse survey translated the explorers' ordinary navigation into a map of the coast. The traverse method had great potential for error because of the problem in logging the ship's track precisely and in getting accurate compass readings on the moving ship. Also, to an observer on a ship, it is often impossible to tell if a feature is an inlet, strait, river, or island lying between the ship and the mainland. A third method, a running survey, involved more time and used the path of the ship as a baseline for triangulation, i.e., measuring angles to objects on shore from both ends of the baseline of travel. The baseline plus angles at each end to the same object provides more accurate location. The choice of method depended on the degree of accuracy needed and the amount of time available to survey the coast.

Captain James Cook followed the practice of going ashore and measuring a baseline on land for greater accuracy using triangulation techniques. He produced his famous map of Newfoundland in 1775 with theodolite and telescopic quadrant, neither of which can be used on a moving ship. However, triangulation is expensive and time-consuming, demanding many trained men, whereas traverse surveys may be done by a single person equipped to find latitude and longitude along his route and to transcribe them onto charts. However, Cook also made excellent maps using the traverse method for most of his voyages in the Pacific.

The quadrant, so named because it was one quarter of a circle, dates back to the thirteenth century as an instrument for astronomers (figure 5.8). By the middle of the fifteenth century, the quadrant was also in use for measuring celestial angles aboard ships. The user sighted a star through the tube, or pinhole, and observed the location of the plumb line on the graduated arc. For sighting the sun, the observer could save his eyes by standing with his side toward the sun and holding the quadrant so that the sunlight passing through the peephole in the front vane fell on the peephole of the rear vane. Thus the instrument was properly aligned to the sun, and the position of the plumb line on the arc could be noted. In the early stages, the arc of the quadrant, like the cross-staff, was marked with destinations

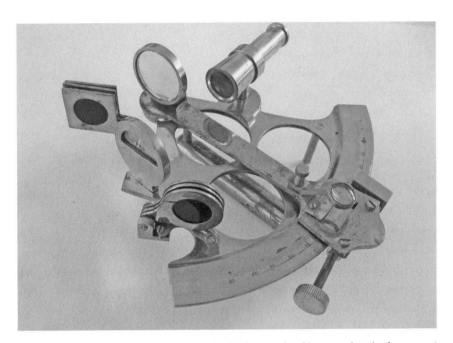

FIGURE 5.8 The quadrant was derived from the astrolabe. The line of sight was along one edge of the instrument and the plumb bob string indicated the vertical angle of the sighted object. Wikimedia Commons. Drawn by Michael Daly.

FIGURE 5.9 When it became necessary to measure both vertical and horizontal angles for computing lunar distances, the sextant was developed in the early eighteenth century. The sextant is more precise and can measure angles greater than ninety degrees. With its movable filters the sextnat may be used to measure directly into the sun. Its use continued through most of the twentieth century, and navigators should still be familiar with it. Photo by E. S. McCoy.

rather than degrees. This allowed the navigator to sail until he reached the latitude of his intended destination, then run along the latitude for the rest of the voyage. Variations on the quadrant—including octants and sextants—gradually became more sophisticated with sun filters, mirrors, and sliding scales graduated in degrees. The sextant continued to be used for navigation well into the twentieth century until inertial guidance and satellite positioning systems appeared (figure 5.9).

Distance of travel was a crucial piece of information for a seaman. He needed distance to estimate position and progress toward his destination. The only means available was "dead reckoning"—estimating the speed of the ship over a period of time as measured on an hourglass. Early navigators could make reasonable guesses of their ships' speed by the feel of its movement through the water. Earliest estimates were made by pacing the deck alongside a small object thrown overboard

FIGURE 5.10 The chip log as a means for estimating the speed of a ship was an improvement over the log or board originally used in the fifteenth century. The line was knotted at intervals and the number of knots that passed the rail in thirty seconds provided a means of computing knots per hour. Hence, the term "knots" for the speed of a ship. Drawn by R. M. McCoy.

and judging the rate of walking. During the sixteenth century, the English made great improvements in navigation techniques and in the instruction of seamen. The log and line for measuring a ship's speed was first among the new developments, first described in 1574.

A piece of wood, originally a log, was tied to a long knotted cord and thrown overboard beyond the dead water at the stern of the ship. The log would remain roughly in place while the vessel moved away. A sailor measured the amount of knotted cord that paid out in thirty seconds, using a small sandglass, as the ship moved away from the log—clearly a two-man job. They counted the small uniformly spaced knots to measure the length of cord paid out in thirty seconds. This could be converted to an estimate of speed as knots per hour. This procedure was repeated anytime the ship changed speed due to a change of wind or change of course. The log was eventually modified to a small triangular board, called a chip log, weighted and tied to stand upright and stationary in the water as the line paid out (figure 5.10). This method of measuring speed was a big improvement over former guesswork, but it was also subject to error because of currents or cord stretching.

The speed and direction information was first pegged on a traverse board, transferred to the logbook, and used for estimating distance traveled. The traverse board had a thirty-two point compass rose drawn on its face with radial rows of eight holes along each of the thirty-two compass points (figure 5.11). For each half hour of a four-hour watch, the helmsman inserted a peg in a hole of the compass direction they were traveling, beginning with the innermost hole and working outward. At the bottom of the traverse board, four rows of holes were pegged to indicate the speed traveled during the hour, with a row of holes for each hour of the watch. At the end of a four-hour watch, all this direction and speed information was recorded in the logbook and used to compute "distance made good."

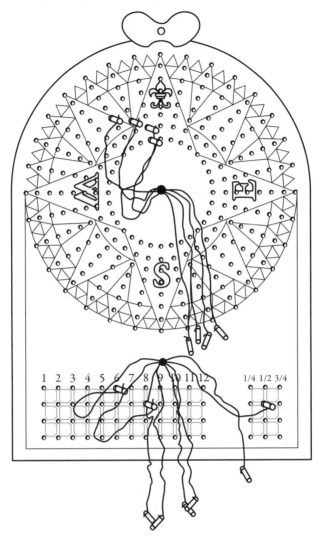

FIGURE 5.11 The traverse board provided illiterate helmsmen a means of recording the direction sailed for each half hour, and the speed for each hour of his four hour watch. Wikimedia Commons.

Because a ship sailing into the wind must tack back and forth rather than maintain a straight course to its destination, a navigator had to keep track of their actual location as well as their progress along the intended course, called "distance made good." By plotting their location on the actual line of sail and drawing a perpendicular connection from that point to the line of the intended course, the distance made good is determined (figure 5.12).

Although astronomers knew methods of computing longitude by measuring the angle between the moon and certain stars, known as the lunar distance method, these techniques could not be applied accurately at sea. Computing longitude by

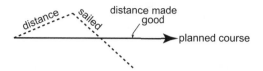

FIGURE 5.12 Sailing toward the wind required frequent course changes, or tacking, resulting in greater distances than if a single course could be maintained throughout. The mariner made a simple plot of actual course to determine the distance gained along the intended course, or distance made good. Drawn by R. M. McCoy.

astronomical methods required much more training and education than the early seaman had, and it required large telescopes on a platform more stable than a ship. The navigator, therefore, had to estimate longitude by keeping track of the east or west distance from a known longitude. Information available to the navigator told him there are approximately 39 nautical miles (44 statute miles) in one degree of longitude at fifty degrees latitude. Therefore, a navigator sailing along the fiftieth parallel to North America would add one degree of west longitude for every 39 nautical miles of westward travel. Considering the many sources of error in estimating distance traveled, it is no surprise that they usually had the longitude wrong—often by a significant amount.

The relationship between longitude and the steady turning of Earth at fifteen degrees per hour was well known, and by the early sixteenth century, some navigators knew that a trustworthy clock at sea would solve the problem of finding longitude. If a navigator had a reliable clock, a chronometer, set to Greenwich time, he could determine his local time by the sun and find the exact time difference. Fifteen degrees for each hour of time difference allowed easy computation of longitude. But it was more than two centuries later before John Harrison, an English carpenter and clockmaker, undertook the challenge to create a ship's clock with the needed accuracy. Many leading scientists of the eighteenth century still believed that a clock of such accuracy could never be made, so they pushed for greater application of the astronomical angles method of time telling. Finally, in 1761, Harrison's efforts produced a clock that passed a prescribed test by accurately determining longitude during a voyage. The clock evolved through several stages from a cumbersome machine the size of a large mantle clock to one that looked like an oversized pocket watch. With each iteration, the design and accuracy improved. Unfortunately for Harrison, the judges were primarily proponents of the astronomical approach, and they denied him the £20,000 prize money because the success was deemed to have been just good luck. Eventually the voyage was repeated, and Harrison created new versions of the clock—a total of five—before he finally received the last of the prize money more than a decade later in 1775, the year before he died.

There was usually a long lapse of time between the invention of any new instrument or technique and its implementation on ships. Resistance to change is always present, and the change to the marine chronometer, as it came to be called, was no exception. Considering how long mariners had been waiting for the solution to longitude, adoption of Harrison's clock was surprisingly slow. Part of the reason was that in the beginning, the marine chronometer was so expensive that most shipowners, unconvinced that it would truly work, were unwilling to buy one. Acceptance of innovation required that navigators become convinced that the new approach was better than their old method, and then persuading owners of the ships to supply the necessary equipment. As often happens, older men retired or died, and their younger replacements accepted the new ideas. Strong support for the marine chronometer came when such a well-known figure as Captain James Cook used a copy of Harrison's clock on his second and third voyages to the Pacific (over a period from 1772 to 1779) and gave high praise for the advantage of having such a clock on board. By the 1830s, although Harrison's chronometer was thoroughly tested and proven, still most navies had them on fewer than a quarter of their ships. Nineteenth-century ships leaving London via the Thames River paused within sight of the Greenwich observatory to set their chronometers with the time ball that fell at 1:00 p.m. daily. In the early 1920s, the ball drop was superseded by a radio time signal, which in turn has been replaced by the atomic clocks in the Global Positioning Satellite System. These innovations for finding position changed traditional navigation methods that had been used for centuries.

6

Martin Frobisher Succumbs to Gold Fever, 1576, 1577, 1578

AFTER THE DEATH of King Henry VII in 1509, England's foreign interests turned back to Europe. His son, Henry VIII, married and divorced Catherine of Aragon, then broke with the Church of Rome. During most of his reign, Henry had threats of invasion from both France and Spain for his flagrant behavior toward the Catholic Church. Therefore, no explorations of the New World occurred during his life. This hiatus continued after his death in 1547 and well into the reign of Elizabeth I, which began in 1558. Even though she had frequent threats from Spain, Elizabeth took a strong interest in exploring the New World for the possibility of a trade route to Asia. By the end of this new stage of exploration, the map of North America had progressed to the state shown in figure 6.1. The coastal map extended past Labrador into Hudson Bay and parts of Baffin Island and Greenland. One of the first Elizabethan seamen in this stage of exploration of North America was Martin Frobisher.

Martin Frobisher (ca. 1535–1594), like most seamen of his time, began a life at sea while still a boy. Both of his parents had died by the time he was 10 years old, and in 1549 young Martin had to move from his home in Yorkshire, near Leeds, to live with his mother's affluent brother, Sir John Yorke. His uncle was a successful merchant and also held a prestigious position as an officer of the Tower mint in London. Merchant John Yorke invested in trading voyages, and soon young Martin was signed onto one of the voyages sailing to Africa. Martin learned quickly, and he soon became skilled in most aspects of seafaring.

FIGURE 6.1 By 1632, eighty-nine years after the map shown in figure 1.1, the map of North America had progressed on the east coast, but the west coast had not changed.

Though Frobisher's uncle gave him rapid entry to life at sea, he did little for Martin's formal education and social refinement. As a result, Frobisher spent his life as a semi-literate who always felt uncomfortable among educated gentlemen and business men. John Yorke apparently felt little obligation to introduce his nephew to the sort of society he himself enjoyed. His primary duty as guardian was to make sure the boy learned a trade, apparently not in his own business. Nevertheless, Martin Frobisher rapidly learned seamanship, and by the age of 20 became commander of a voyage to Africa. During this period he became acquainted with Michael Lok, another merchant and seafaring traveler, who later became the patron for Frobisher's voyages to seek the northwest route to Asia around North America.

For a period of time before his voyages of exploration, Frobisher became a privateer sanctioned by Queen Elizabeth. Frobisher showed an unfortunate tendency to attack and plunder ships of England's allies as well as its enemy, Spain. This practice landed him in English jails on several occasions, but few other details are known of his adventures during this time.

By 1520, Magellan had discovered a southern route around the seemingly impassable barrier of South America. Frobisher, like John Cabot and many others, thought it only logical that there must also be a passage to the north. Frobisher perceived that discovery of such a passage was the only thing in the world not yet done that could bring him instant fame and fortune. He sought out his acquaintance, Michael Lok, about funding a voyage of discovery to the North American Arctic.

Michael Lok was a mercer involved in foreign trade of textiles. His father had been the personal cloth merchant for King Henry VIII. Lok was also the London agent of the Muscovy Company, formed to supervise trade with Russia and to establish trade routes across the north coast of Russia reaching Cathay and India. In this role, Lok naturally had a great interest in trade routes, maps, and the promise of foreign places. The Russian route had been tried by this time and proven impossibly icebound. So Frobisher's proposal was exactly what Michael Lok wanted to hear.

With the permission of the Muscovy Company, which already had a charter for trade with Asia, Lok and Frobisher offered a proposal to the Queen's Privy Council for a voyage to discover a northwest route to Cathay (China). The Council gave its approval, and Lok and Frobisher formed a new company called the Cathay Company; they found additional investors, and the voyage was on. Prominent among the investors were the Earl of Warwick and some members of the Queen's Privy Council.

Despite all efforts to raise funds, the trip was meagerly provided. The three ships available were small and in poor condition. The funds (£875) could support only thirty-five men for the three ships, making them dangerously undermanned. Two of the ships were barks, the *Michael* and the *Gabriel*, at thirty tons displacement each. The third ship was a pinnace of only ten tons and was not identified by name. By comparison, Cabot's ship, nearly eighty years earlier, was fifty tons displacement. Figure 6.2 shows the route for Frobisher's third voyage; the previous two voyages were the same except for the brief entrance into Hudson Strait made in the third voyage.

Michael Lok went on the first voyage, but the primary written account of the voyage is one by George Best, who was among the officers and gentlemen aboard the *Gabriel*.[1] The objective of the first voyage was to expand trade.

First: The great hope to fynde our English seas open into the seas of East India by that way [northwest] . . . whereby we might have passage by sea to those rich cuntries for traffik of merchandize, which was the thing I chiefly desyred. Secondly: I was assured by manifolde good proofs of dyvers travailers and histories that the cuntries of Baccaleaw,[2] Canada, and the new fownd lands thereto adjoining, were full of people and full of such commodities and merchandize, as

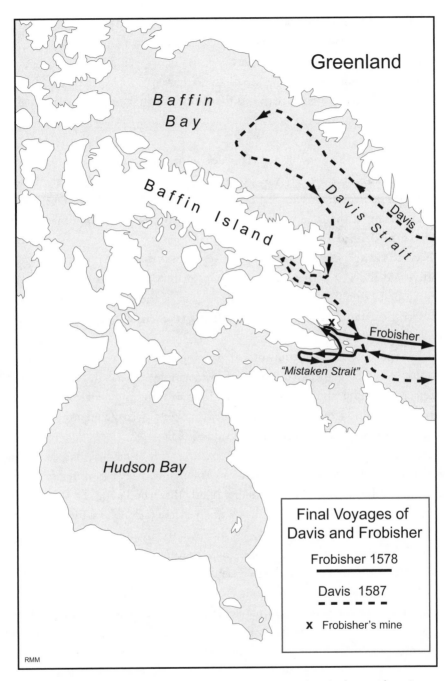

FIGURE 6.2 Routes of Martin Frobisher and John Davis, master mariners in the queen's service.

are found in the cuntries of Lappia [Lapland], Russia, and Muscovia [Grand Duchy of Muscovia, including the city of Moscow], . . . which are furres, hydes, wax, tallow, oyle, and other. Whereby yf yt should happen those new lands to stretch to the north pole so that we could not have passage by sea that way, . . . yet in those same new lands to the northwestward might be established the like trade of merchandise . . .[3]

Lok appeared to have established a fall-back position in case Frobisher's logic about finding a passage in the north was wrong. If northwest trade routes to the Orient failed, then perhaps the Arctic areas themselves would be fruitful for commerce. The general feeling at the time was that many productive lands without masters still waited to be claimed for profit.

Because the purpose of Frobisher's first voyage was the discovery of a passage to Cathay and India, it was essential that maps of new lands and routes be made. George Best created one of the first maps made of the Arctic regions of North America. Best stated, "I applied myselfe wholy to the science of cosmographie, and secrets of navigation."

To that purpose, many instruments for mapping were purchased for the voyage. A brief list of purchases included: a book of cosmography, a metal globe, an armillary sphere, a brass level, and a wood cross-staff for measuring angles of stars above the horizon. In addition, they took the latest maps available, twenty compasses, eighteen hour glasses, and a brass astrolabe for measuring angles.

The list of supplies also included an interesting array of medications, consisting mostly of herbal compounds and minerals. Among the many useful items brought on board, Frobisher loaded coal, thinking rightly that wood might be scarce in the far north. Also, he took three hogsheads of aquavitae (alcoholic spirits, e.g., whiskey) primarily for the officers and other gentlemen, and five tuns of beer for the crew. Frobisher purchased his own mattress and bedding from company funds, but common sailors took their own blankets and slept on the deck.

Although maps were made following Frobisher's voyages to North America, only a sanitized version of them became public. The English desire to establish supremacy in commerce with Asia made maps of newly discovered lands and routes the objects of highest secrecy. In the public account of Frobisher's voyages the printer stated that, "albeit I have in a fewe places somewhat altered from my copie, and wronged therby the authoure, and have soughte to conceale upon good causes some secretes not fitte to be published or revealed to the world (as the degrees of longitude and latitude, the distance, and true position of places, and the variation of the compasse)."[4] However, Bynnyman, the printer, assured the reader that the map was sufficient to make the narrative of the voyage easily understood. This urge for secrecy, so important to national interests, had a downside. Voyagers from other countries could unknowingly claim the same territory unless the first claimant had started a settlement or left a monument. This was not always done.

THE FIRST VOYAGE, 1576

Frobisher's three ships began the voyage to America in mid-June 1576, with a farewell wave from Queen Elizabeth I as they sailed down the Thames past her palace in Greenwich and turned north toward the Shetland Islands. By the end of June, a week-long storm made progress impossible. When the storm finally passed, the pinnace was missing along with its four-man crew and was never seen again. The two remaining ships continued, and in a few days they sighted an unknown land, which they could not approach because of the numerous large icebergs. Frobisher thought he had found the legendary island of Friseland, but in fact, this "unknown land" was the first reported sighting of Greenland since the middle of the fifteenth century, about 130 years before Frobisher, when the Norse abandoned their settlements on the southwest coast. The earlier Greenland experience of the Norse voyagers was unknown to the English.

At this point, the crew of the *Michael* had lost heart for discovering new lands and decided to head back to London. It appears that the threatening sight of the icebergs and the rugged coast of Greenland gave them pause. Frobisher, in consultation with men on the *Gabriel*, decided to continue the expedition. When the *Michael* reached London, the crew reported the face-saving lie that Frobisher's ship was lost. That lie very nearly came true.

On the day after the *Michael* left, the *Gabriel* was hit by another severe storm and turned flat onto its side. Being open at the waist, the *Gabriel* quickly began to fill with water. The ship was clearly going to sink, and the crew was in a state of despair. According to Lok's account, Frobisher courageously saved the day by ordering the men to cut the mizzenmast (the aft mast) to lighten the weight, and the ship slowly righted itself. It was full of water, and many supplies had washed overboard, but they were afloat. After riding out the remainder of the storm, the crew pumped the *Gabriel* dry and rigged a new mizzen mast. Frobisher was still determined to continue westward rather than face the humiliation of returning to London in failure.

At the end of July, the crew of the *Gabriel* saw an island with a prominent headland, which they named Elizabeth Foreland, now called Resolution Island, located off the southeast end of Baffin Island. From there they sailed into a long narrow waterway—about 35 miles wide at the mouth—presumed to be a strait that could be the passage to Asia. Their supposed strait seemed to have an eastward flowing current, which they took as confirmation that this waterway was open to the west and would form a passage to Cathay and India. Frobisher convinced himself that the land on the north side of "Frobishers Streytes" was Asia, and that the highlands on the south were in North America. Today we know they were in a 150-mile long bay, called Frobisher Bay, at the south end of Baffin Island. Not until the mid-nineteenth century did a ship, headed by Charles Hall, finally sail Frobisher's waterway far enough to determine that it was actually a bay rather than a strait.

On August 19, they went ashore for the first time on an island near the "strait's" north shore. Climbing to the top of a high hill to get their bearings, they saw seven strange-looking one-man boats (kayaks) heading for the island. They hurried back to their own boat and rowed out to their ship, ready to defend the ship if necessary. Frobisher sent a boat crew back to the shore to try to converse with the unknown men. The Eskimos (Inuits) appeared to be friendly, so the crew induced one of them to come on board while one of the ship's crew stayed on land as a pledge of sincere intentions. Best reported that the visitor on board was impressed by everything he saw. The crew gave him cooked meat and wine, but he appeared to care for neither. They gave him some trifles that attracted him and took him back to shore. However, Frobisher felt uneasy about trusting the Eskimos and sailed to a safer position on the south shore. He was unwilling to risk defending his ship with a crew of only eighteen men, some of whom were weakened from disease, probably scurvy.

They saw some men paddling a single large boat (likely an umiak, which is larger than a kayak and will hold several men). The Eskimos made some sign of friendly intentions, but Frobisher, still feeling vulnerable, indicated that one of them should come on board alone. In the usual hostage trading gesture, the ship's master went ashore, and the Eskimos took him to their camp.

From his position on the ship, Frobisher thought the Eskimos on shore were trying to signal their companion aboard ship to jump into the water and escape. Frobisher suddenly took a long dagger and held it to the chest of the hostage, threatening to kill him if he moved. The ship's master continued with the Eskimos to their encampment and saw how they were living, which he described as "strange and beastly."[5] When the master was returned to the ship, the Eskimo hostage was returned to the land and another Eskimo volunteered to go aboard the ship. The crew gave him "many tryfles of haberdash" (clothing items) which greatly pleased him. Frobisher, through signs and gestures, persuaded this Eskimo to pilot them through the "strait" to the West Sea (Pacific Ocean) toward Cathay. The man was to go ahead paddling his kayak and the ship would follow. They offered him many items which they laid out for him to see. The man agreed and indicated that it would take only two days of rowing. Still not trusting the Eskimos, Frobisher sent five crewmen to row the Eskimo ashore to a designated spot within view of the ship so Frobisher could rescue them if anything should go wrong.

The five men dropped the Eskimo at the designated spot then rowed out of sight against Frobisher's order, despite shouts for them to return to the ship. Neither the five men nor the Eskimo was ever seen again by those on the ship, nor was their fate ever learned. Frobisher tried repeatedly to find out if they had been captured. Though Frobisher concluded they had been taken by force, some authors have suggested it was possible that the five men intentionally deserted ship to live with the Eskimos to escape the stern discipline Frobisher maintained on ship. The captain had thereby lost five more able-bodied men and the ship's boat as well. He remained in the area through that day and night hoping to see the men again.

The next morning (August 20), Frobisher set sail in the supposed strait. As he passed near the shore within sight of the Eskimo settlement, he had the trumpet sounded and shots fired into the air, but no one was in sight. The next morning he returned to the place where he had last seen his men and saw nothing, but heard voices and laughter from Eskimos on the shore. After three days, he returned once more, but the site was still deserted, and he was ready to give up ever finding his men and boat.

Frobisher despaired that he would be unable to carry out his mission without a full crew and the only available boat needed for going ashore. By this time, he had only thirteen men and boys. This was a serious loss. He lacked enough men to handle or defend the ship efficiently, and without the small boat he could not go ashore. Without any physical evidence to verify his discoveries, he would face great humiliation. He would rather face death than be exposed to that humiliation.

Frobisher prepared the crew to defend the ship if necessary—always a captain's first consideration. He had canvas nailed down to cover any chains, shroud lines, or anything that an outsider might grab to climb onto the ship, which rode low and near the water level. He made a plan to shoot and sink one of the Eskimo's big boats, some of which held as many as twenty men, and take prisoners as hostages to recover his own men.

Frobisher was pleased to see a number of Eskimos paddling kayaks and umiaks toward his ship and prepared the reduced crew to defend their ship in the hope of capturing some of the Eskimos. As they approached the ship, the Eskimos stopped when they saw men mustering in the ship's deck and a small cannon set up against them. Their decision, wisely, was to move farther away and to the other side of the ship, away from the cannon. They apparently had some idea of its purpose.[6]

One man approached paddling a kayak and making signs of friendship. Frobisher returned his friendly signs and motioned for him to come closer. Frobisher stood alone at the waist of the ship where it is nearest the water, with firearms at his feet that the Eskimo could not see. Both parties gave the appearance of friendliness with no hint of ill intentions. Frobisher offered small items to the man as inducements, but the man was very wary and would not come closer. The captain threw a shirt into the sea along with other items that would float, and the man in the kayak retrieved them.

Frobisher held out a small bell, which of course would not float, and rang the bell to entice the man to come close enough to reach for it. As the kayak came closer one of the sailors tried to snag it with a boathook. But the Eskimo saw what was happening, pushed away from the ship and would not come back. Frobisher was sorely frustrated and tried to lure him back with the bell again. Eventually the Eskimo returned to the ship cautiously, ready to push off again if alarmed by anything. Frobisher tossed the bell intending it to fall short and drop into the sea. The Eskimo's obvious disappointment encouraged Frobisher to try again. He held another bell in view, but not with a fully extended arm, so the man would have to come closer. When the man finally reached for the bell Frobisher grabbed him by the wrist and quickly lifted man and kayak together into the ship.

Frobisher, without showing any enmity toward the man, made signs to let him know that he could go free if the five sailors were returned. Not surprisingly, the Eskimo seemed not to understand Frobisher's meaning. All the other Eskimos could see these events and departed quickly with much shouting. The captain waited through the next day to see if the Eskimos would return his men.

The time had come for more serious thought about the course of action. Frobisher had lost five men. Freezing nights had already begun, and they were not prepared to stay over the winter. The crew were feeling tired and sick from the hard labor of the voyage. Frobisher conferred with his men to decide on their course of action. Their conclusion was that to continue onward with winter approaching would be a threat to the entire expedition, and all might be lost in the end.

On August 25, they set sail homeward, with the Eskimo still on board, and arrived in England on October 9—forty-five days of travel compared to the sixty-one stormy headwind days on the trip out. They met several storms on the return voyage, but the wind was in their favor and gave them good time. In London they were greeted with great enthusiasm and admiration by the people who were fascinated by the "strange man" and his boat. No one had ever seen Eskimos before.

This voyage marked the first recorded encounter of Europeans with the Inuits since the end of Norse contact in Greenland in the fifteenth century, about 130 years earlier. Most English people had no awareness of those earlier contacts, so this was truly a new discovery for them. Unfortunately, the Inuit man they brought home had caught a cold during the voyage and, with no resistance to this illness, died a few months after reaching England—but not before being put on display to many crowds of people.

THE BLACK ROCK

When Frobisher's ship had first reached Frobisher Bay, a party of men had been sent ashore to bring back anything they found as tokens of possession. With these articles, Frobisher then claimed the land, named *Meta Incognita*, as a possession of the queen. One of the items found and brought aboard was a large black stone about the size of a loaf of bread.

Frobisher had promised Michael Lok that he would give him the first thing he found in the new land. That item was this black rock found on the shore of the "strait." Lok, being an enterprising man, determined to see if the rock had any commercial value. He broke off pieces which he took to three different assayers for evaluation. All three concluded it was marcasite, a disulfide of iron very similar to pyrite, and having no value. Lok, just to be certain, took another piece to a fourth person, Giovanni Agnello, a Venetian living in London, who claimed to find some gold in it. Lok gave this man two more pieces of the rock from which Agnello again produced a small amount of gold. Lok became convinced that this fourth assayer was correct—a good example of selective belief. When he asked Agnello how he could find gold when the other three could not, Agnello said, "Nature needs to be flattered," suggesting that he had special methods. Agnello no doubt felt that customers needed to be flattered too.

This cryptic remark was apparently enough to convince Lok that Agnello had some special skills and that his hopes for gold had come true. He hastened to write to the queen, whom he knew would be interested in funding another voyage if there was a possibility of gold. "Moste humbly I crave pardon in troublinge your majesty with the readynge of this wrytynge,"[7] he began. Lok explained in detail all the results of assays by four men, one of whom found gold in the rock on repeated assays. We must give Lok credit for being forthcoming and honest, for deceiving the queen could be his ruin. Lok's letter even quoted Agnello's remark about flattering nature. He made clear that Agnello had sworn he was being honest in his assessment and that it was entirely true. He told the queen that Agnello had even asked to have some of the ore for himself, but Lok refused him, saying that it came from the new land and that the company of investors in the expedition had all rights to the ore. However, Lok said he had made Agnello give a promise of secrecy concerning the value of the ore. Lok further related that even when Frobisher had asked him about the outcome of the assays, Lok told him that the assays found nothing of value but a little tin and silver in the rock. Michael Lok's letter informed the queen that her own state secretary, Sir Francis Walsingham, believed Agnello to be an alchemist and that Walsingham had ordered additional assays. These new assays showed some tin but no gold. Lok's letter revealed that Agnello had proposed that the two of them sail secretly to *Meta Incognita* to pick up a load of the ore for themselves, but Lok had explained that such a venture would be illegal. Lok must have been trying to squelch any possible rumor that he was trying to defraud the queen at the risk of his own life. Lok heard a rumor that he had leaked the secret about finding gold in the ore. In his letter to the queen, Lok strongly denied that this was true, and swore to yield his goods and his life to her Majesty if it were proven that he had told the secret. Poor Michael Lok was gripped by fear that he would offend the queen or the Privy Council, which was headed by Walsingham, and find himself in mortal trouble. Yet the prospect of untold wealth urged him on.

England had watched Spain and Portugal haul loads of wealth back from their claimed lands in the Americas. Although England had acquired a good share of the New World treasures by hijacking on the seas, here was England's opportunity to get some gold at the source. The prospect of gold from the Americas seemed to have infected the queen and her Council.

Elizabeth decided to invest £1,000 in a second voyage. When she supported the venture, how could others decide not to go along? Even the skeptic Walsingham put in £200, along with a total of £5,150 from thirty-eight Cathay Company investors, many of them from the Privy Council. This was an enormous amount of money compared with the total investment of £875 for the first voyage. The irresistible scent of gold had cast its spell.

THE SECOND VOYAGE, 1577

The instructions from the queen and other investors for the second voyage put ore mining at top priority. This stipulation essentially precluded finding a new route to the Orient. After gold, the next priority was to search for the five missing men from the first voyage. The probability of doing all those tasks and still exploring for Frobisher's presumed passage was remote. Further, Frobisher was cautioned not to let the search for the passage take so long that he would be forced to stay over the winter. Given the short ice-free sailing season in the Arctic, investors expected to have the ore back in England by the end of summer. To assure that the expedition could not waste too much time searching for the Passage, they were ordered to send only one of the small ships no more than 100 leagues (about 300 miles) to explore the "strait."

Frobisher began the second voyage on May 25, 1577 with three ships. Two were the barks—*Michael*, the bark that turned back on the first voyage, and the *Gabriel*— that had completed the first voyage. The third was the *Ayde*, a tall ship of two hundred tons to carry a large load of ore. A total of 120 men, including officers, sailors, soldiers, and thirty miners, manned the three ships. The *Ayde* was the command ship carrying a hundred men. The remaining men were divided between the two smaller ships.

Hardtack, flour, pickled beef and pork, dried cod fish, butter, oatmeal, honey, rice, cooking oil, and vinegar were among the food provisions taken on board. Eighty tuns (256 gallons per tun) of beer provided a ration of one gallon per day for every man. Five tuns of sack and malmsey wines were loaded for the officers and other gentlemen. These supplies were intended to provision a voyage of six months.

Upon arrival in the Frobisher "Strait," they immediately began searching the various islands and found several with a good supply of the ore. When they found signs that Eskimos were in the area, Frobisher sent about forty crewmen ashore to establish contact. Frobisher and representatives from each group came together to exchange goods and information.

Frobisher's priority again was to take hostages to give him some bargaining power for the return of the five missing Englishmen. When he reached out to grab hostages, the icy ground made it impossible to hold them. The Eskimos ran to some bows and arrows they had hidden nearby and began shooting at Frobisher, who was shot in the buttock with an arrow as he was trying to escape back to the ship. The crewmen on board heard the shouting and began firing on the Eskimos. Frobisher was able to capture one of them and take him to the ship. During this onshore melee, the ship's kitchen caught fire, creating some onboard excitement. At the same time, ice floes moved in closer and threatened to damage the ship. The sailors had to cut anchor quickly and move away from the ice to avoid damaging the ship. This was not a wonderful beginning, nor was it the best way to gain cooperation. It marked the end of peaceful meetings between Frobisher's expedition and the local population. Future encounters omitted the pretense of negotiation and trading and went straight to attempted kidnapping or shooting.

In a few days they established a base on an island about 45 miles in from the mouth of the "strait." It had a good supply of the ore and a good harbor protected from ice floes. They named the island Countess of Warwick for one of the major investors in the voyage. The present name is Kodlunarn Island, an Inuit word meaning "white men's land."

On the mainland, sailors found an unoccupied Eskimo settlement site, and by rummaging through the tents discovered some items of English-style clothing, a canvas vest with many holes that could have been made by arrows, a shirt, a girdle, and three shoes of various sizes. They concluded the items belonged to their five lost men. Thinking there might be a chance the missing men were still alive, they left a written note. They also left paper, pen, and ink so the men could write an answer if they saw the note. Returning later, they found that the Eskimos had removed all trace of the settlement. In a later encounter, another battle followed as the English fired their guns and the Eskimos shot their arrows. The Eskimos were ultimately overwhelmed by the English, and surviving Eskimos jumped from the rocks into the sea rather than face capture. The sailors succeeded in capturing an Eskimo woman with an infant and took them back to the ship.[8]

The next time the Englishmen found a group of Eskimos, they promised to return the captured man, woman, and child to them if they would return the five missing sailors. The Eskimos indicated the men were still living, and they could take a note to them and return in three days. Frobisher sent a letter to the missing men explaining that he would do everything possible to get them back either by barter or by force, and that he had hostages to offer in trade. Moreover, he issued a threat that if the Eskimos failed to deliver, he would not leave a single hostage alive. He sent pen, ink, and paper with the Eskimos so the men could respond.

Four days later the Eskimos returned. Frobisher went over to them with high expectations. As he approached in his boat he became aware that many Eskimos were hidden behind rocks out of view from the ship. Frobisher surmised that the Eskimos were hoping to capture someone in order to redeem their own hostages from the ship. So the Englishmen stayed away from the shore, but noticing some of the Eskimos creeping behind the rocks toward them, Frobisher gave up and returned to the ship without any news. He never discovered the fate of the men.[9] The Eskimos made other attempts to entice Frobisher's men ashore, but each time a potential ambush was evident, forcing them to return to the ship. Frobisher now turned his attention to the primary task of mining ore from Countess of Warwick Island.

On August 20, their mining work was done. They had gathered two hundred tons of supposed gold ore in twenty days with only five miners and the help of some of the "gentlemen" and soldiers. The men were tired, some were badly injured, their shoes and clothes were worn out, and their ore baskets and mining tools were broken, but the ships were full. Also, they noticed that ice was forming around their ship at night, giving good indication of the lateness of the season. They dismantled tents, made bonfires, and fired a volley of shots in honor of the Lady Anne, Countess of Warwick. On August 24, 1577, they set sail and left *Meta Incognita* behind. That night six inches of snow fell on the deck of their ship.

Upon receiving samples of the ore, assayers in England produced a wide range of estimates on the amount of gold in the ore. Several assayers found no sign of gold, and some promised they could deliver ten ounces of gold per hundred pounds of ore, which would have been very profitable. This wide discrepancy led to the assayers criticizing one another's methods and equipment. Michael Lok wrote a letter to investors in which he explained that the assayers could not agree, but he felt that the ore was very rich. Lok estimated that they would earn a profit of £40 per ton of ore. This outcome convinced the queen to support a third voyage to the site of the mines with enough ships to collect a generous amount of ore. That settled the question of disputes among the assayers for the present, and the investors of the Cathay Company were suddenly committed to investment in another voyage without actually extracting any gold from the existing ore. The investor's response to the variable assay results was that a larger volume of ore would be needed, and therefore, more ships should be sent on the expedition. The second voyage had cost the Cathay Company

£6,410. As the investors had provided only £2,500, the deficit made everyone desperately want the two hundred tons of rock to be gold ore. This partly explains why they were so willing to accept the positive assays rather than the negative results of other assayers. The third expedition began without anyone having refined a single ounce of ore from the second voyage.

THE THIRD VOYAGE, 1578

The new plan was to send four or five ships and collect eight hundred tons of ore—four times the amount taken on the previous trip. Thinking that more ore would make bigger profits, the plan soon expanded to fifteen ships capable of carrying two thousand tons of ore. The expedition would leave one hundred miners in *Meta Incognita* with enough supplies for the winter and following summer. Three ships would stay for the miners' safety in case they should need to leave before the supply ships returned the next year. Then the arriving ships would either retrieve the men or leave them supplies for another year. The planners were unaware that the sea in the *Meta Incognita* area was frozen from late November until July, and the miners would not be able to leave during that time, even in an emergency. Nevertheless, it was sound planning to provide escape for the miners in case supply ships never returned from England.

Before they left, the fifteen captains went to the queen's court in Greenwich to receive her best wishes. She gave Frobisher a gold chain and presented other gifts to all the captains. Before embarking, Frobisher gave instructions to all captains on the course they should sail and on keeping order in the ships. Rule number one was no swearing, dice playing, card playing, or filthy talk. Religious services and prayers were to be held twice daily. They all set sail for *Meta Incognita* on May 31, 1578.

On July 2 they again saw the promontory they had named Queen's Foreland, but were diverted from the "strait" of Frobisher's Bay because of ice. The summer of 1578 was a much colder summer than they had experienced on the two previous voyages. There was a very old belief that saltwater could not freeze, but fresh water in bays and inlets would freeze and float out to the open sea. It was believed that salt, as well as tidal action, would prevent the open sea from freezing. Another factor contributing to this belief was that ice even 100 miles from land tasted fresh when melted.[10]

When the ships tried to make passage, ice floes moved about with the wind and tide, and openings appeared long enough to let one ship through, but closed before a second ship could enter. This required much rapid taking in of sails to prevent collision with large ice floes. The result was frequent separation of the ships as they tried to move about.

George Best wrote of one incident in which a ship was crushed by ice floes. "And one of our fleete named the barke *Dennys*, being of an hundreth tunne burden, seeking a way amongst these ise, received such a blowe with a rocke of ise, that she sunke downe therewith, in the sight of the whole fleete. Howbeit, having signified hir daunger by shooting of a peece of great ordinaunce, newe succour of other shippes came so readily unto them, that the men were all saved with boates."[11] Unfortunately, the *Dennys* contained the materials for a house to be used by the men staying the winter in *Meta Incognita*.

By now, the entire fleet was surrounded by ice, having progressed westward. Then a great storm arose, putting them in great danger by closing the openings in the sea ice behind them. Some of the ships hung cables, beds, masts, and planks over the sides to protect the ship from the ice. Some ships anchored to the lee side of a floe and hoped for the best. A few ships were lifted in the water by ice pressing on both sides. Despite all precautions to protect the ships, the hull timbers of some were crushed.

After a few days, Frobisher determined that they were on the north side of a strait, but could not distinguish terrain features in the fog and mists. Eventually he realized their latitude was too far south, and they were in the wrong strait. The battle with ice floes had taken their ships off course, and they were now in the strait that later became known as Hudson Strait, though they simply referred to it as the "Mistaken Strait." They had sailed about 180 miles into the unfamiliar strait, losing twenty days of valuable time. In the Mistaken Strait they could see signs of a fruitful land with much more grass and game than in their intended Frobisher Strait farther north. Also, they saw and traded with some Eskimos in this new area.

Why did they not take more interest in this strait? They obviously knew this was a different strait and believed there were continental lands on both sides. If the expedition's objective had still been discovery of a northwest passage, they probably would have explored it further and would have become the discoverers of what is now Hudson Bay. The urgency of bringing home the ore convinced them that they had spent enough time away from their task and true destination, so they turned around and headed for Frobisher Strait.

The crews of the various ships resisted continuing through the ice, and some said they would rather face hanging for mutiny than to perish in the horrible icy seas. Many urged stopping in a harbor until the wind could clear the ice away, giving them a chance to refresh themselves and repair damage to the ships. However, they pushed ahead into a storm, and the ice pressed all around them as they sailed out of the Hudson Strait. The snow was so thick they could scarcely see each other nor open their eyes to handle ropes and sails. Their clothes became soaked. But they continued through whenever they saw a gap in the ice, and all ships eventually found their way into Frobisher Strait.

On July 31 they reached Countess of Warwick Island, where one of the mines would be. In addition, they mined other islands where the ore had been found. They quickly started mining operations as an open pit excavation and established a temporary settlement for miners. Masons built a stone house, which they finished by the end of August. However, because most building materials had been lost when the *Dennys* sank, miners would not be able to stay over the winter as planned. Other issues influenced the decision against anyone staying for the winter— primarily the miners feared that snows in summer must portend an unsurvivable winter. The deciding factor came when Frobisher saw a particularly brilliant display of the aurora borealis, which he took as a warning that they should leave.

Within a month after their arrival at Countess of Warwick Island, the ships were fully loaded and ready for the voyage home. The transfer of bags of ore from beach to ship was made during heavy storms, and several ships were blown away from shore, inflicting some damage to every ship. They departed, or escaped, on August 31.[12]

As they approached the British Isles, another great storm threatened the fleet, and one ship was grounded and abandoned on the west coast of Ireland. The others arrived in England October 1, 1578. The expedition returned with 1,350 tons of ore and forty fewer men. The ore was locked away in Bristol and in a castle along the Thames River near Dartford, awaiting construction of smelting furnaces.

THE OUTCOME

These expeditions provided a burst of enthusiasm for England to have a source of gold in America, as did the Spanish and Portuguese. However, the efforts to extract gold, now begun in earnest, ended in absolute failure. The final assays found nothing of value. The investors quickly turned bitter, and Michael Lok, treasurer for the Cathay Company, was accused of dishonesty, with Frobisher himself among the accusers. Michael Lok furiously accused Frobisher of bringing back valuable ore on the first voyage, but worthless ore on other trips. Expenses of building furnaces and further assays forced Lok to assess the stockholders an additional £9,270 for the heavy expenses. At the end of five years, the total investment finally reached £20,160 with no return.

Lok himself had invested £2,200 of his own money, the loss of which ruined him. He went to debtors prison on at least eight different occasions. His fellow investors abandoned him to the mercy of the creditors. Investors refused to pay pledged amounts to cover expenses already incurred, including pay for sailors and miners. Lok became the focus of creditor's efforts to collect, and he was bankrupted, sued, and imprisoned. His name and signature was on all the transactions and documents in support of the voyages, assays, and furnaces, making him the legally responsible person. Lok later claimed to have seen the inside of every London jail. He was released for the last time in 1581, spending the next years trying to put his life back together and rebuilding his finances and reputation. On top of it all, Lok, who had been married twice, had a total of fifteen children and stepchildren to support. As late as 1615—thirty-seven years after the last voyage—Lok was sued for a debt of £200 resulting from the northwest ventures.

Frobisher's part in causing the queen to lose £4,000 in the venture explains why he was not given any significant commands for the next few years. His reputation was badly damaged, but he was not hurt permanently. He remained inactive and essentially unemployed with no obvious source of income before regaining his good standing and command of a ship. In 1585 he was given a vice-admiralty under Francis Drake for a raid to the West Indies. His main glory came with his success in the battle against the Spanish Armada. He was given command of the largest ship in the operation and was highly successful in achieving a major victory in the engagement. For this he was knighted on board the fleet's flagship, the *Ark Royal*. After this, he continued to receive naval commands.

Frobisher married twice. He owned several manor houses by the time he died in 1594 at the age of 55. He died of a gunshot wound during an attack on a Spanish fort at Brest, France.

Recent analysis and assay of the infamous ore shows it to be high in iron, aluminum, chromium, and nickel. The rocks contained the mineral marcasite, which is an iron sulfide and has the same chemical composition—but a different crystal structure—as pyrite (fool's gold). Gold is present only in trace amounts. High gold content found by some assayers in 1578 may have been due to the crude state of the art of assaying, incompetence, or more likely, due to secretly salting the furnace with gold during analysis.

Upon returning to England, George Best produced a full account of the three voyages. One of his important contributions was a map of the world showing details of a newly discovered supposed passage to Asia, "Frobisher Straits." Unfortunately, Best's additions to Ortelius's map are no help in determining where the expedition actually went. Best's map is a modification of Ortelius's world map that was completed as part of a world atlas in 1564. Best overlooked mentioning Ortelius as a source. Another map showing only the area of the voyage is more help, but still provides more questions than answers (figure 6.3). On this map, Frobisher Strait is shown passing between North America and Greenland. Later maps showing Greenland in its correct position also show Frobisher Strait at the south tip of Greenland. Is this due to the printer Bynnyman's concealment or Frobisher's error as some have proposed, or is it simply an error introduced by other mapmakers? The land area called "Greenland" may have been shifted from south Baffin Island to present day Greenland by mapmakers and

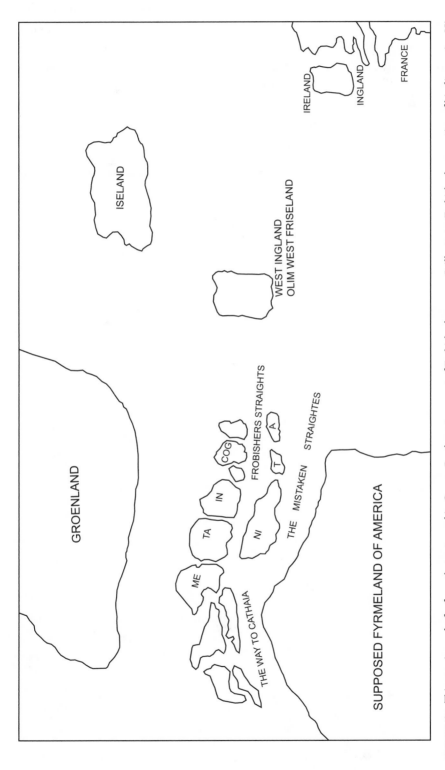

FIGURE 6.3   This map is a redraft of one that appeared in George Best's 1578 account of Frobisher's voyages. It illustrates Frobisher's perception of his discoveries. His sighting of Greenland is shown as West Friseland, and the name Greenland is applied to Baffin Island. The chain of islands is his Meta Incognita.

the Frobisher Strait mistakenly moved with it. Frobisher was able to return to his "strait" at Baffin Island twice with little problem, suggesting that his estimation of longitude, though probably not accurate, was not the source of such a gross error. Frobisher's misadventures led to discoveries and maps that proved to be mistaken, or at least overblown, but the transformation of a quest for the Passage into a frenzy for gold made his a worthwhile story.

7

John Davis Makes a Near Miss, 1585, 1586, 1587

LITTLE IS KNOWN about John Davis (c. 1550–1605) during his early years, except that he was born in Devonshire around 1550. However, his skills as a mariner caught the attention of some investors in the Muscovy Company when they invited him to lead an expedition in search of the Northwest Passage. The Muscovy Company was formed by a group of merchants in 1551 under a charter from the king for the purpose of establishing trade with Russia and finding a northeast passage to China. The northeast passage had proven impassable, but the company continued a monopoly on trade with Russia until 1668. After finding the northeast route blocked by ice, only two possible routes to Asia remained for the English—over the North Pole or along a route north of North America. The Spanish and Portuguese already controlled the southeast and the southwest routes.

In 1585, Queen Elizabeth I granted the Company a patent authorizing them to sail north, northeast, and northwest with as many ships as they could muster. They had full rights to a trade monopoly with any newly discovered lands, and they would become the governing authority in any colony they might establish. Under this directive, John Davis began the first of three voyages.

These voyages established Davis's reputation as a careful and able navigator. In his role as navigator, Davis made a great contribution to the art of navigation in 1609 by writing an instruction book for mariners, *The Seaman's Secrets*.[1] He wrote that the minimum need for successful navigation must include the compass, the cross-staff, and a chart. In the relatively unknown area where Davis was headed, the chart available to him would have shown the limited discoveries of Cabot and Frobisher, along with the basic grid lines showing latitude and longitude on which to plot his route. Davis acknowledged in his writings that the astrolabe and the quadrant provided unreliable information at sea. The swinging plumb bob of the quadrant provided little chance for accuracy on a rolling ship. In *The Seaman's Secrets*, Davis wrote directions for using the cross-staff plus his own invention, the back-staff, by sighting the sun or the North Star. He also established the routine of keeping data from the traverse board in a book that became the ship's log, with columns for compass heading, time traveled on that heading, distance made good, solar observations of latitude, and wind direction. With all his attention to proper procedures, John Davis made a great contribution to the standardization of navigation techniques.

FIRST VOYAGE, 1585

In June 1585, Davis set sail with two ships, the fifty-ton bark *Sunneshine* and the thirty-five-ton bark *Mooneshine*. Davis himself captained the *Sunneshine*, and the entire expedition carried a company of forty-two men. Refer to Chapter 6, figure 6.2 for a map of his voyages. By mid-July, they reached the east coast of Greenland amid dense fog and drifting ice and "the most deformed, rocky and mountainous land that we ever saw."[2] About 3 miles of land-locked sea ice separated them from the shore and prevented them from sailing closer. Davis gave it the name Land of Desolation. John Janes, a member of the crew who wrote a narrative of the voyage, noted that their breakfast that morning was a half pound of bread and a pint of beer.

On August 1, 1585, the two ships crossed a stretch between Greenland and Baffin Island, which became Davis Strait, and made landfall at an inlet that Davis named Exeter Sound for the city in Devon that sponsored him. Many of the landmarks in this area have retained the names given by Davis. From this point, Davis turned southward along the coast, and in a glow of satisfaction with his success so far, Davis made a generous increase in food rations to all crewmen.

A short distance south on Baffin Island, Davis named a promontory Cape of God's Mercy, today's Cape Mercy. Davis soon entered an inlet, which he named Cumberland Sound, for an investor in Davis's voyage, but he failed to push all the way. He took soundings and found the water getting deeper as he sailed farther into the sound. Also, he reported seeing whales and strong tidal currents coming from the west. These promising indicators convinced Davis that he was in a passage to the Pacific Ocean. Nearing the head of the sound, Davis became concerned with a change in the weather and felt they might be frozen in for the winter if they continued. Having provisions enough for only six months, he beat a retreat. They left Cumberland Sound for England on August 24. Once again an explorer had fallen short of finding the inevitable negative answer to the question of a passage, and had hopefully assumed he had found it. This seemed to be requisite for funding another voyage. The summer of 1585 may have been unusually warm, as the expedition had met no ice barrier in Davis Strait nor in Cumberland Sound. Also, he met no Eskimos on Baffin Island; perhaps the tales of Frobisher's kidnappings and killings ten years earlier kept the Eskimos out of sight.

SECOND VOYAGE, 1586

Davis gave a glowing report to his investors and to the queen's first secretary, Sir Francis Walsingham. He told them that the Northwest Passage was a certainty, that it was deep and ice free. The merchants quickly bought into a second and larger voyage with four ships, including the *Sunneshine* and the *Mooneshine*, plus the 120-ton *Mermayd* and the small ten-ton *North Starre*. On May 7, 1586, they departed. One month later, two ships, the *Sunneshine* and the *North Starre*, separated from the others and headed north to examine the possibility of a route passing over the North Pole. This side venture failed because of ice before they reached the Arctic Circle. During their return to England, the little *North Starre* became lost in a storm and was never seen again.

Meanwhile Davis continued slowly westward. He sailed around the south end of Greenland and up the west coast as far as 64° N latitude and anchored in a bay, probably at the location of Nuuk (formerly Godthåb, dating back to the Norsemen). Immediately, swarms of kayaks came to greet them, and the Englishmen stayed in the attractive harbor for several days enjoying the mounds of fish and fowl the Eskimos brought to them. They had music, dancing, and sporting contests (jumping and wrestling) with the Eskimos, finding them to be nimble jumpers and strong wrestlers. Here the crew assembled a small pinnace that had been stowed on one of the ships, and it became a third member of the expedition.

The friendly relationship with the Eskimos turned sour when the Eskimos began taking items from the ship. The Englishmen fired cannons to frighten them away, but the next day Eskimos returned, and the stealing began anew. When the English sailors again drove them away, the Eskimos began throwing rocks at the ship, injuring some of the seamen. By now it was July 17, and it was clearly time to move on after spending two weeks of the brief summer sailing time in the bay.

As Davis continued north along the Greenland coast, the weather turned bad and the water had many icebergs. The sails and masts became encased in ice and the men pleaded to return home. Davis compromised by sending the *Mermayd* home with about half the company, then continued with the *Mooneshine* and the pinnace across Davis Strait. The point where Davis made landfall was the eastern point of the Cumberland Peninsula, and his reckoning for distance and latitude position was within a few miles of the correct location. Davis repeatedly showed this degree of navigational accuracy in his voyages. Partly this accuracy should be attributed to his invention of the back-staff, but much credit goes to his concern for careful readings. For this he gained a reputation as a master navigator.

At this point Davis turned south to return toward Cumberland Sound, however for unknown reasons, he sailed right past Cumberland Sound. He even rediscovered his Cape of God's Mercy without recognizing that he had been there. The visibility may have been very poor due to a storm, causing Davis to focus on keeping the ship safe and steady in the high winds, or perhaps his position determination was wrong. The next day, the ships were farther south near Hall's Island, which Frobisher had visited. Davis was too far out to see the opening to either Frobisher Bay or Hudson Strait, and he continued south about halfway along the coast of Labrador, sailing into a bay now known as Davis Inlet. There he stayed for several days, replenishing water and hunting partridge and pheasant.

On September 1, he again headed south and stopped near an inlet, now called Hamilton Inlet, and thought it looked promising as a passage to China. Although he could not enter the inlet because of high head winds, he must have realized that he had passed Cumberland Sound. As Davis continued along the coast, he reported catching many codfish and going ashore to cure them for the return voyage. Another storm, this time from the northeast, threatened to ground his ship on the shore, but they were saved by a timely change of wind. This marked the end of his second voyage, and he set sail for England on September 11. Stormy weather had played a decisive role in this fruitless voyage. Davis faced returning to his investors with nothing positive to report. Undaunted, he put a positive face on a lame situation by saying he had discovered at least four possible routes that should be explored. Davis took a chance and offered to make the next voyage without charge to the investors. He intended to finance the voyage by selling his share of his family's estate, assuring all investors of a profit.

THIRD VOYAGE, 1587

John Davis must have convinced his supporters to sign on for a third voyage. It is not clear how much Davis spent of his own money, or if he actually sold his share of the family estate. However, we know he sailed on May 19, 1587, with an expedition of three ships, the familiar *Sunneshine* of fifty tons and two smaller ships, the *Elizabeth* and the *Ellen*.

First landfall was again on the west coast of Greenland near Nuuk (Godthåb). The *Sunneshine* had sprung a leak that required constant pumping. Apparently many of the crew had been induced to sign on with the promise of making the voyage a combination exploration and a profit-making fishing expedition. The sailors soon became eager to get on with the lucrative fishing part of the venture and forget about exploration. Davis compromised by allowing part of the crew to take the *Sunneshine* and the *Elizabeth* fishing, while Davis took the little pinnace *Ellen* exploring. This generous act by Davis was prompted when the crew became agitated and appeared close to mutiny if fishing were postponed until the end of the voyage.

On June 21, the ships parted company, with Davis heading north along the Greenland coast. He continued up the coast as far as latitude 72° 46′ N in the vicinity of present day Upernavik. This would have been a fortunate latitude for crossing Baffin Bay, because the true entry to the Northwest Passage, Lancaster Sound, lay nearly due west.

Crossing Baffin Bay was not a simple task. A persistent ice pack filled much of the center of Baffin Bay for the entire year. It diminished in the summer and expanded in the winter, but for many centuries it never disappeared. Davis attempted to push through the ice with a combination of sail and oars and soon found himself surrounded by ice. Luckily, he managed to retreat and return to open water, but still on the east side of the ice pack. He sailed southward to the southern edge of the ice pack and then sailed west to reach Baffin Island near Exeter Sound, which was by now familiar territory to him. On July 19, Davis reached the mouth of Cumberland Sound and felt he was again on the verge of finding the passage. On the first voyage he had found Cumberland Sound and used its potential as a basis for the second trip. On the second voyage, he had sailed past and missed it altogether. Now on the third voyage, he sailed all the way to the head of the supposed passage and discovered a dead end.[3]

While there, he noted a magnetic variation of 30° west. Davis was one of the earliest mariners to keep a record of magnetic variation across the Atlantic, and he was among the first to note an increasing westerly variation. Other mariners continued taking magnetic readings, and by the nineteenth century, mariners began to realize there must be a north magnetic pole different from the geographic north pole. Sir James Ross discovered its location on the west side of the Boothia Peninsula in 1831, however the location drifts with time, and the magnetic pole is now nearly 800 miles farther north and continues to drift slowly. This changing data was invaluable to mariners for making adjustments on their navigation compasses, and accurate magnetic readings were essential to their mapping efforts.

Davis next passed by an inlet he named Lumley's Inlet for an important man in the queen's court. Lumley's Inlet was in fact Frobisher Bay, where Martin Frobisher had wasted so much time collecting useless ore. At the time of Davis's voyage, Frobisher's strait was still erroneously shown on maps as cutting through the southern part of Greenland. This error persisted into the nineteenth century.

Sailing southward past Frobisher's *Meta Incognita* on August 1, Davis passed the opening of a great gulf with waters swirling and roaring like the meeting of two tides. He called this turbulence a "furious overfall," which he believed was caused by the coming together of two strong currents, creating a turbulent and strong flow eastward. The gulf he saw was no doubt the opening to Hudson Strait, and entering the strait would have led him to the discovery of Hudson Bay, not as exciting as China, but a great discovery nevertheless. The plausible reasons Davis did not turn west were that ice floes may have been gushing along with the current, or perhaps the current was simply too strong for his small ship. He mentioned a mass of ice flowing out the mouth of the strait. Also, Davis may have felt an obligation to make his rendezvous with the half of his expedition that went fishing in the Grand Banks. He was to meet them between 54° and 53° on the coast of Labrador. Davis waited there for a reasonable time, passing the days with hunting, and then set a course for England on August 15. One month later they arrived in their home port of Dartmouth and found that the *Sunneshine* and *Elizabeth* had already arrived, loaded with fish. Although the explorations yielded nothing of value, the bounty of fish convinced the fishermen of Dartmouth to begin their voyages to the Grand Banks of Newfoundland.

John Davis had found nothing of interest to his investors, however he paid them the compliment of permanently adding their names to many features of the landscape. To his great credit, Davis also added several thousand miles of newly charted coastline to the emerging maps of the Arctic along western Greenland and eastern Baffin Island.

8

Henry Hudson Has a Very Bad Day, 1607, 1608, 1609, 1610

HENRY HUDSON (1570?–1611), English sea captain, had tried to sail to China across the North Pole in 1607, hoping to encounter an open polar sea. Ice prevented his success, but hope for an open polar sea persisted for more than two centuries. Hudson had tried to reach Asia the next year by sailing east along the north coast of Russia, and failed again because of ice. He made two unsuccessful attempts to find a sea route through North America in 1609 and 1610. At one time, King James I had Hudson restrained from leaving England because he had claimed New World discoveries for the Dutch. Sailing under contract for another friendly country was accepted practice for a successful navigator, but King James was not amused that such desirable lands as New Amsterdam should become Dutch rather than English. Although Hudson was a reputable navigator and explorer, he sometimes met strong resistance from and had trouble with his crews, which came near mutiny more than once.

Given these experiences, Hudson might be considered a failure, both as an explorer and as master of his ship. But passing judgment on Hudson is not that straightforward. Hudson was a man of great vision and imagination who saw possibilities for new sea routes and wanted to test them himself. He accomplished many things during his four attempts to find a route to China. He sailed about 140 miles up the river in New York that now bears his name and wrote that the land was rich and fertile for agriculture. This news caused a surge of interest in colonization and settlement by the Dutch. He found an immense bay that now carries his name, along with the strait that enters the bay. Because he did not explore the western shores of Hudson Bay, the world was led to think he had found the elusive passage to Asia. This belief led to further expeditions by other explorers. On his first voyage, Hudson reported an abundance of whales at the group of islands called Svalbard (once called Spitzbergen). Soon whaling in the area was a valuable boost to the English economy. All these events elevated Hudson among his countrymen. Even the king was persuaded to release Hudson from his travel restriction so he could conduct another voyage in search of a northwest route to the Orient.

Almost nothing is known of Hudson before he took his first command to explore a polar route to China. He may have been an employee of the Muscovy Company and perhaps learned his seamanship under their guidance. No certain birth date or birth place is known, his level of education is unknown, and no information exists about his sailing experience prior to 1607. Hudson suddenly appears prominently on the scene in 1607, and his sudden disappearance in 1611 greatly elevated public interest in his voyages. There is a possibility he that sailed with John Davis in 1587, but that story may have been a fabrication written to satisfy the public's curiosity about Hudson.[1]

FIRST VOYAGE, 1607

Supported by the Muscovy Company of England, in April 1607 Hudson began his first known voyage as ship's master to search for a route over the north pole to the Orient in the small bark, *Hopeful*. He set out on May 1 with eleven crewmen, including his young son John, and sailed northward to slightly beyond 81° north latitude (less than 600 nautical miles from the pole) before he was stopped by pack ice and turned back. The most useful contribution of this voyage was to report a great number of whales near the north shores of Svalbard, a group of islands over 500 nautical miles north from Norway. See figure 8.1 for routes of Hudson's voyages to North America.

Hudson brought the *Hopeful* back to London in September. The push for a route over the North Pole was prompted by a common belief at the time that the polar seas would be free of ice, though no one had yet been there. Two misconceptions supported this belief: first, that saltwater in open seas, well away from the fresh waters near land, would not freeze, and second, that the twenty-four hours of sunlight during polar summers would provide enough heat to prevent ice formation. Hudson's encounter with ice should have disproved these ideas.

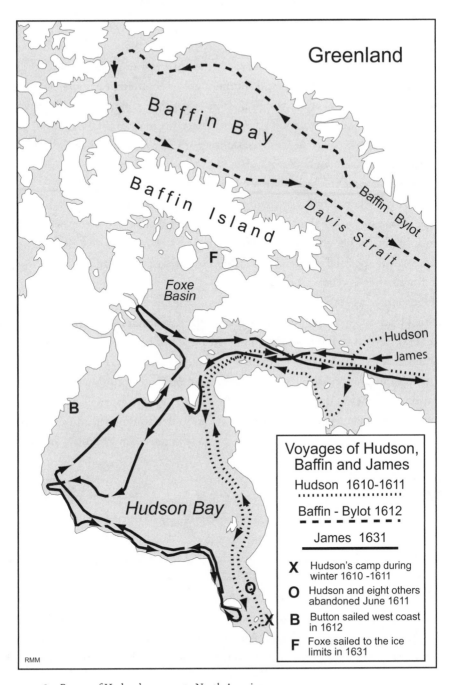

FIGURE 8.1 Routes of Hudson's voyages to North America.

SECOND VOYAGE, 1608

The Muscovy Company decided next to try the northeast route. In April 1608 they sent Hudson with the *Hopeful* and fourteen sailors to attempt a route across the north coast of Russia, even though at least two other explorers, Sir Hugh Willoughby and Stephen Borough, had already tried this route and failed. The voyage was uneventful except for the reported sighting of a mermaid by two of the sailors as reported in Hudson's logbook. ". . . shee was come close to the ship's side, looking earnestly on the men: a little after, a sea came and overturned her: from the navill upward, her backe and breasts were like a woman's, as they say that saw her; her body as big as one of us; her skin very white; and long hair hanging down behinde, of colour blacke; in her going down they saw her tayle, which was like the tayle of a porposse, and speckled like a macrell."[2] Perhaps they saw a large albino seal in foggy conditions after a night with the keg.

Hudson got as far as Novaya Zemlya, an island off the north coast of Russia, before being stopped once again by ice, and concluded that this route was impassable. He returned to England in August. The crew had become agitated when they learned that Hudson might extend the voyage and possibly head for North America. Protests from the crew may have pressured Hudson to abandon that idea, though Hudson made it clear that he returned to England of his own free will. This incident, though not violent, was Hudson's first near mutiny. By this time, the Muscovy Company was losing interest in exploration—possibly because the trip to the northeast was so fruitless.

THIRD VOYAGE, 1609

The Dutch East India Company became interested in finding a trade route to the Orient, and they invited Hudson to try the northeast passage again. Aware of Hudson's passion for finding a northwest passage, the Company explicitly ordered that he was not to attempt a northwestern route in North America, even if he failed to find a northeast passage. Hudson apparently had no intention of going east, despite the written restriction on his mission. His experiences in the first two voyages had rightly convinced him that it would be futile.

Hudson left with a crew of sixteen in the 120-ton *Half Moon* in March 1609, heading dutifully northeastward up the coast of Norway. At North Cape, where he would begin the easterly leg of the trip, he wrote in his log that unfavorable winds made progress difficult. He therefore gave up without further effort and headed west for North America—his own intended destination. According to *Half Moon* crewman Robert Juet's intermittent journal entries, the turn around could not have been because of the weather. His journal showed fair weather and favorable winds for that time.

On June 12, the *Half Moon* reached what is now Nova Scotia and headed south along the American coast, at one point raiding and robbing an Indian village, supposedly to prevent the Indians from attacking. They entered Chesapeake Bay and Delaware Bay without landings and turned back northward. In early September, reaching New York Bay, they stopped for a time and had friendly encounters with the Indians.

Hudson sailed up the river as far as the site of Albany, about 140 miles. On the return trip down river, they had a hostile encounter with Indians and killed several of them. This prompted them to begin the return trip home without further exploration, returning to England in November 1609. Hudson was detained and forced to stay in England, but sent the ship and the Dutch component of the crew on to the Netherlands. King James I would have no more of this sailing for the Dutch by his leading explorer.

The ship's log records the Indian name for the large island in the lower reach of the river as Manna-Hata. The Dutch returned with settlers and established New Amsterdam at the mouth of the Hudson River, creating the first beginnings of New York City. This site remained with the Dutch East India Company, but was ceded by treaty to the English in 1674 after negotiations based on John Cabot's sweeping claim of vast areas of land in 1497.

FOURTH VOYAGE, 1610

Now, for the first time, Hudson had explicit directions from a group of investors, merchants, gentlemen, and members of parliament to explore his long anticipated Northwest Passage. Hudson had eliminated any possibility of a passage across the North Pole, or eastward toward Russia, or below 60° latitude in North America. This left one route to try—northwest. His plan was to find the "furious overfall" mentioned in logbooks by John Davis in 1587. The backers for his fourth voyage were wealthy Englishmen John Wolstenholm, Sir Thomas Smyth, and Sir Dudley Diggs. Hudson no longer represented the Muscovy Company or the Dutch East India Company.

In May 1610, Hudson sailed down the Thames with a crew of twenty-two, including his son John, listed as "a boy." Five of the crew had sailed with Hudson on previous voyages. One was the first mate, Robert Juet, who had earlier shown his ability to foment discontent among the men. Hudson might have used more caution in his choice of crew.

The *Discovery* sailed northward along the east coast of Great Britain, past the Isle of Orkney and the Faeroes. Near the end of June, they arrived at the eastern opening of the strait (Hudson Strait) where John Davis had reported the "furious overfall"[3] in 1587. At last Hudson was at the site he fervently believed to be the gateway to the Orient. Hudson felt confidant that the current and turbulence of the overfall was caused by tidal flow from a distant ocean. In fact, the size of Hudson Bay is sufficient to create a strong tidal flow though Hudson Strait into the Atlantic Ocean.

The ship proceeded into the strait against the current, skirting south of the ice floes. The crew had never sailed among large floes and icebergs, and they became very apprehensive, especially when they witnessed a large iceberg capsize. This event impressed on them the dangers of sailing in these icy seas. The next day a great storm forced them to retreat to a leeward position among the icebergs until the storm ended. As the *Discovery* continued westward, it became completely enclosed by ice, presenting a new source of anxiety for the crew. Such a situation can strand a ship for months, or worse, crush and sink it. Fortunately the ice soon reopened, and the frightened crew continued sailing west.

Now their position measurements showed that they had sailed 300 miles farther west at this latitude than any other European. Hudson asked the crew if they wanted to go on or not. Far from feeling pleased with the accomplishment, some of the crew had had enough of sailing in the Arctic ice and wanted to head home. Others supported Hudson and wanted to continue. Heated arguments erupted among the men.

The decision, of course, lay with ship's master, Henry Hudson, and he predictably ordered a continuation of the expedition. But the very act of asking the men their opinion perhaps opened the door of dissension for the crew. It may have lowered their confidence in the judgment and authority of Hudson. Sailors below the level of first mate would have been unaccustomed to an officer asking for their opinion.

At the point where Hudson Strait opens into the vastness of Hudson Bay, the expedition anchored and went ashore to hunt for game. In that area, Hudson named and mapped two promontories, Cape Wolstonholm and Cape Digges, for two of the expedition's sponsors. The name Digges applies in that area today to two islands north of the settlement of Ivujivik, Q uebec.

The men who went ashore on this treeless tundra found many wild fowl and an abundance of grass. They also found great quantities of scurvy grass, known for its ability to prevent or cure scurvy. Scurvy grass is not truly a grass, but is a salt-tolerant, spicy-tasting plant in the cabbage family, often found near coastal areas in the Arctic. In the same area, the sailors found quantities of sorrel, which also contains high quantities of vitamin C needed for curing scurvy. Although these men knew nothing about vitamins, they knew that certain plants could prevent scurvy.

The shore party found unoccupied skin huts that they likened to haystacks. On the inside they discovered stores of game birds hanging on the walls. They naturally took the birds and carried them back to the shallop, the small single-masted boat that was towed behind the *Discovery*. Rather than linger in this bountiful place to take advantage of the fresh food supplies, Hudson ordered that they set sail southward along the east shore of the great bay collecting data for mapping the shoreline.

After several days they came into an embayment, now called James Bay, at the south end of Hudson Bay. It became clear there was no waterway through the land and they were at a dead end. Some arguments again arose among the men over the value of staying in this southerly bay rather than exploring farther west in hopes of finding a passage to a western ocean. They were into September, and signs of winter were apparent.

On September 10, the rebellious first mate Robert Juet called the entire ship's company together for a meeting, with the intention of exposing the "abuses and slanders" that Hudson had made against him. The meeting backfired against Juet, however, when crewmen testified to Juet's overwhelming abuses and mutinous actions against Hudson. One man said that early in the expedition, Juet had predicted that murder and bloodshed would occur on the voyage. Another said that Juet had threatened to take the ship and return home, but that Hudson had convinced him to calm down. A third said that Juet had persuaded some of the crew that they should keep swords and loaded muskets in their quarters. Juet had warned them that the muskets would be used before the voyage was over. These and other similar testimonies made it apparent that Juet was dangerous to the expedition and that he must be punished. Hudson removed Juet as mate and replaced him with Robert Bylot, who was experienced and well respected by the crew. Hudson also promised that he would forget Juet's offenses if he behaved for the rest of the voyage. Juet was allowed to remain on deck as a regular seaman, which did little to suppress his rebellious impulses against Hudson. Juet had a persuasive and frightening effect on the more timid crewmen. The long stay in James Bay for no apparent reason, along with the prospect of wintering there, had already brought some of the crew to the point of mutiny.

Following the confrontation with Juet, Hudson continued sailing within James Bay, presumably searching for a suitable wintering spot. They sailed into the southern extremity of James Bay in the last days of October, when the nights were already long and the ground was covered with snow. On November 10, the ship became locked in ice for the winter. The *Discovery* had begun the voyage in May with no plan for spending the winter away, meaning they had expected to be home no later than November. Now November had come, and the ship was icebound for the winter. Careful rationing of the remaining food was crucial. The only resupply up to this time had been at Cape Digges, where they had acquired a good amount of fowl and scurvy grass. Hudson resorted to rewarding anyone able to kill any game or fish.

A young man named Henry Greene was one of the crewmen on board the *Discovery*. Little is known about him, but he apparently had been taken, for unknown reasons, into Henry Hudson's household some time before the voyage. Greene's name was not on the crew roster provided to the sponsors at the time of sailing from London, but he was taken on board secretly at Gravesend near the mouth of the Thames. From the beginning, Henry Greene was another disruptive presence in the life of the crew. He was argumentative with certain men, and soon the whole crew disliked him. When some complained to Hudson about Greene, Hudson advised them to let the matter drop. Hudson's support for Greene was an opportunity for first mate Juet to turn up the heat a bit. He told members of the crew that Hudson had brought Greene aboard to spy on potential troublemakers. Unfortunately, Hudson did nothing to refute the accusation against Henry Greene, and Greene remained under suspicion by the crew. When Hudson first heard of this skullduggery, the *Discovery* was only a day past Iceland, its last port of call. He would have been wise to return to Iceland and put Juet ashore where he could find passage to England on a fishing boat. But Hudson must have felt that Juet's experience as a seaman was too valuable to lose and may have may lacked confidence in his own navigation skills.

Hudson's perceived favoritism toward Greene again created friction following the death of one of the crewmen, John Williams. The dead man's belongings were brought on deck for auction, a common practice at the time. Greene pled with Hudson in advance of the auction for a particular item of clothing, and Hudson decreed that Greene should be the one to buy the garment. This affront to the accustomed practice of seamen further alienated Hudson from many of the crew.

Hostility escalated following an argument between Hudson and the ship's carpenter, Philip Staffe. The carpenter refused to follow Hudson's order to go ashore to build a house for winter quarters. Staffe complained that the snow and cold was such that he could not work, and furthermore, he was a ship's carpenter and not a house carpenter. The next day, the carpenter took Henry Greene and went ashore to hunt fowl. Hudson became angry that Greene would befriend the belligerent carpenter and decided to give the dead man's garment to Bylot instead of Greene. When Greene returned, he protested so vehemently that Hudson decreed that Greene's wages were to be cut off entirely. Thus did the ship's company settle into a hostile and dangerous state of mind for the long winter ahead. The infuriated Greene became active in stirring up dissent against Hudson.

The meager supply of provisions was amply enlarged by an abundance of ptarmigan in the area. One crewman, Abacuk Pricket, wrote that they killed 1,200 fowl in three months' time. For those months, each man aboard could eat about one-half a bird per day. When the ptarmigans migrated, they were replaced by swans, geese, and diverse kinds of ducks. Without the fortunate supply of meat for the winter, the crew no doubt would have starved. They made it through most of the winter, but when the supply of birds ended, many weeks remained before they could sail again. As the ice began to break up, the men began to fish and reported catching five hundred fish on the first day. Later attempts were less successful, and soon they were able to catch only eighty small fish in two days. The food situation was again in crisis. Hudson set out in the ship's shallop to search the coast for signs of humans with whom they could barter for meat. He found none who would show themselves, though he saw evidence of their presence.

Now Hudson came to the point of distributing the last of the food. As they prepared for the homeward voyage, Hudson distributed all the remaining bread, which came to one pound per man. Then he divided five cheeses equally among the men. Some of the men were unable to regulate their eating and finished their allotted rations in a day. A breaking point came when Hudson ordered a search of the crew's personal sea chests for the possibility of hoarded food. The man assigned to gather the hidden food brought thirty cakes of bread in a bag. A few days later, Henry Greene became the primary voice of the plotters' plan to force Hudson, his closest supporters, and any sick men into the ship's boat and leave them behind. Greene rationale was that the ship had fourteen days of provisions remaining, and only with fewer mouths to feed might they make it back to England. The plotters argued that they had not eaten in three days, and Hudson was not showing any intention of getting underway.

The mutineers were resolute. What they had begun they intended to see through to the end. When warned that they would be hanged, the mutineers vowed they would rather hang at home than starve in the icy wilderness. Furthermore, anyone who did not fully support the mutiny would be abandoned with the rest. Robert Juet raged that he would cut the throat of anyone who tried to stop him. He intended to keep the anger at a boil, lest some fainthearted mutineers opt out and leave Greene with too few men to carry off a mutiny. They could easily become the ones set ashore.

Abacuk Prickert, who wrote the only surviving account of the mutiny, claimed that he tried to delay the revolt until he could personally negotiate with Hudson concerning the grievances of the men. Prickert said he was alarmed that Greene's intent was revenge and bloodshed. Prickert told the plotters that if they would delay, he would join them and try to justify their actions to the authorities when they got home. When Greene refused to delay, Prickert made Greene and the others swear on a Bible that they would do no harm and that whatever they did was for the good of the voyage.

Juet proved to be more vehement than Greene in his intent to take the ship and said he could easily justify the deed to the authorities when they returned to England. Prickett said that Juet and his followers each came and took the oath to do no harm—as though being marooned in the Arctic was not harmful. Prickett wrote that he was hopeful the company of mutineers would allow Hudson and his supporters to return to the ship in due time. Prickett also said he was prevented from warning Hudson of the impending revolt. At daybreak one day, a sailor came to Prickett to get water for the kettle. As Prickett went to the water barrels in the hold, the mutineers shut the hatch and locked him in.

As soon as Hudson came out of his cabin, three mutineers tied his arms behind his back. Juet attempted to capture John King, who had been made first mate in place of Robert Bylot, but King drew his sword and would have killed Juet if other seamen had not intervened. The shallop was brought to the side of the ship, and nine men embarked, including Hudson, his son John, and first mate John King. The mutineers allowed them to have a musket with powder and shot, several pikes, an iron pot, and some meal. The abandoned men felt sure the mutineers would bring them back to the ship. This was a mistaken hope, because both Juet and Bylot stayed on the ship, and both knew navigation well enough to manage. As the *Discovery* set sail, the tow rope to the shallop was cut, and the small boat was left behind.

The mutineers took the occasion to pillage the castaway's sea chests and belongings. They also ransacked the ship searching for hidden provisions. Their search yielded some meal, two casks of butter, twenty-seven pieces of pork, and half a bushel of peas. In Hudson's cabin they found two hundred biscuits, a peck of meal (one quarter bushel), and a butt of beer (126 gallons). This discovery later helped confirm the mutineer's position that Hudson was hoarding food for himself and a few favored men. In the midst of the looting and searching, a lookout noticed that the shallop was coming in sight and sailing toward them, trying to keep up with the *Discovery*. On this news, the crew set full sail and left the castaways behind forever. The leaders of the mutiny clearly had no intention of rescuing the luckless men in the shallop.

The *Discovery* sailed up the east coast of Hudson Bay in June 1611, stopping once to hunt fowl, until it again became locked in shifting ice floes for two weeks. The crew now recognized Henry Greene as captain of the ship, and Bylot navigated. They began to talk of the reception they could expect in England and even considered the prospect of staying at sea. After stopping twice to hunt game, twice running aground on underwater rocks, and once having a friendly encounter with a group of natives, the crew had a second encounter that turned deadly. Six men went ashore for more trading with the natives and gathering more sorrel. After a good experience on the previous day, they went completely unarmed except for a knife carried by

Prickert. A sudden battle erupted and two crewmen were killed immediately. Others fled to the boat and pushed off toward the ship under a hail of arrows. Greene made it back to the boat with a mortal wound from a fatal arrow. Over the next two days, two more men died from arrow wounds.

This disaster left five men dead and reduced the mutinous crew to only eight. Despite this reduced number, they still lacked enough food for the long voyage across to England. They went ashore again and managed to kill four hundred fowl without another encounter with the natives. Even with this bounty, the crew was rationed to one-half a bird per day. This suggests they were anticipating being home in one hundred days. The *Discovery* at last turned homeward through the channel now called Hudson Strait, entered the open sea, and set a course they hoped would take them to Ireland. Robert Bylot was now master of the ship. They sailed many days without certain knowledge of their location and began to fear they had missed Ireland. Trouble continued to follow them. Robert Juet died unexpectedly. The last fowl was eaten, and the crew gave up hope, no longer doing the necessary work. At this dire time, they at last sighted Ireland, but the people there would give them nothing without money. A few more days of sailing took them to England and the end of their voyage.

INVESTIGATION BY THE HIGH ADMIRALTY COURT

The crew of the *Discovery* had every reason to expect immediate arrest and jail. However, they went free. A month passed before an investigation was made by the Merchant Adventurers who sponsored them. On October 24, 1611, the official report said in part: "They all charge Master Hudson to have stolen the victuals by means of a scuttle or hatch cut between his cabin and the hold; and it appears that he fed his favored companions, such as the surgeon, and kept the others at only ordinary allowance, which led those not so favored to make the attempt and to perform it so violently. But all conclude that, to save some from starving, they were content to put away so many." The report concluded that they all deserved to be hanged. Conveniently for the crewmen, all entries in Hudson's journal after August 3, 1610, had been mysteriously destroyed, and none of the survivors seemed to know who had done the deed. However, the loss of this portion of Hudson's journal did not prevent mapping the east shore of Hudson Bay. Logs showing the ship's location and data for subsequent maps had been kept by the first mate as primary navigator.

Six years passed before any legal action was taken against the men. In 1616, the High Admiralty Court brought charges of murder against the entire crew. The lapse of time allowed the crew to prepare well. Their statements to the court were completely uniform and consistent. They were unanimous in blaming Greene and Juet—those two and others who had died on the return voyage—as the primary perpetrators. Each one could say they took no part in the mutiny and did not know of it until it was already underway. Their main defense was the fear of starvation, and the meager food situation was well documented. To have expressed any doubt of Hudson's competence would have brought no sympathy from the Admiralty Court whose guiding premise is the omnipotence of the ship's master. If incompetence had been their defense, the crew almost certainly would have been imprisoned and hanged.

The mutineers told a tale of peaceful separation that stretches the imagination. The nine men who were put into the shallop with a bag of meal and a little powder and shot for their musket must have known they had no chance of survival, and would not reasonably have gone without protest. All the sailors were acquitted in 1618. Robert Bylot had been pardoned earlier for his skill and seamanship in bringing the ship safely back to England. Was the crew justified in taking such extreme action as mutiny? Did they have reason to distrust Hudson? Did Hudson's initial favoritism for Greene cause distrust? Was Hudson hoarding food for a select few on board? Some details of the situation may help answer these questions.

First, Hudson appeared indecisive and weak in time of conflict. This is unusual behavior for a sea captain who must be decisive and quick to issue punishment for wrongdoing. Although not sufficient cause for mutiny, Hudson's weakness must have lowered the crew's respect for him. Second, Hudson showed little competence as a navigator and depended a great deal on Juet, which may have been a primary reason for keeping Juet in spite of his trouble making.[4] Third, well into the strait, Hudson gave up an essential part of his command by asking for a consensus to continue. Crews of any vessel would regard that as another sign of weakness. Fourth, in James Bay, Hudson sailed back and forth for several weeks looking for some sign of a westward passage. This appeared to the crew as aimless wandering at a time when winter was near. This lost time assured they would have to spend winter on a ship with insufficient provisions. Hudson's vacillation between anger and conciliation toward the crew would have made them feel they were in the hands of an incompetent leader. By the following spring, when Hudson seemed unable to get underway for a return to England, the crew felt the need to take matters into their own hands.[5] This they did with harsh severity.

The map made after Hudson's voyage showed the eastern shore of Hudson Bay, but as no effort was made to explore the western shore, that portion of the map was left blank. To the mapmaker's credit, there was no imaginary passage to the Pacific drawn on the map. Notes of the voyage described a strong current and tide moving through the strait—sufficient to lift the ship off a shoal on which they were once grounded. Such a current was believed to be evidence of a tidal flow between two oceans. The very lack of a western boundary on the map enticed further speculation that the elusive passage lay just beyond.

Robert Bylot came through the ordeal of mutiny and trial unscathed. In fact, the knowledge and experience he gained on the Hudson voyage was to his advantage. He had been in Hudson Bay, though it was still considered a possible passage to the Pacific, and his experience soon earned him command of a return voyage.[6] Investors were eager to continue the search, and Bylot was the man to do it. In 1615 and 1616, Bylot sailed as ship's mate with William Baffin on two return voyages in search of a passage.

The first voyage went through Hudson Strait and north into Foxe Channel, where they were stopped by impassable ice. Baffin concluded that they should have continued farther north in Davis Strait to search for a westward channel. The two men tested this idea in the second voyage the next year. As they sailed up Davis Strait and Baffin Bay, they passed an ice-choked opening, which they named Lancaster Sound, but they failed to recognize it as a passage—the only navigable passage—to the west. Today's Baffin Bay and Baffin Island were both named after William Baffin. Although they explored new waters and coastlines, they failed in their objective of finding a passage. Robert Bylot's name remains today on an island in the Canadian Arctic and a bay in Greenland.

PART III

West from the Pacific; Overland to the Arctic Ocean, 1728–1789

One does not discover new land without consenting to lose sight of the shore for a very long time.

—ANDRE GIDE

9

Bering and Chirikov by Sea, 1741; Hearne, 1770
and Mackenzie, 1789 by Land

A HIATUS OF nearly ninety years elapsed during which no new expeditions from western Europe were sent into unknown territories of North America. Kings of England, France, and Spain each faced internal problems and had no interest or resources for exploring new lands. After Elizabeth's long and prosperous reign, the Stuart and Georgian line of kings proved to be inept at keeping the peace both internationally and internally. Recurring wars with France and Spain, not to mention a civil war in England, kept exploration out of consideration. This lack of expansionist activity eventually came to the notice of the Russian tsar, who wanted to expand Russia's presence in the world. The result was a Russian venture to Alaska under the command of Vitus Bering. At the end of this period of exploration in 1789, the map of North America advanced on the west with little change on the east. Mackenzie and Hearne ventured to establish two points along the vast space between the east and west ends. See figure 9.1.

Vitus Bering (1681–1741), a Dane serving in the Russian navy, sailed in 1728 in command of the *St. Gabriel* under orders from Tsar Peter I. He sailed from the far eastern port of Kamchatka to determine where Asia connected to North America. This effort was part of Tsar Peter the Great's plan of expansion that transformed Russia from an insular nation into a major European power and resulted in a major Russian foothold in North America. Bering sailed through the straits that now have his name, but the fog prevented him from seeing the American shore. Hence, the location of the western extremity of North America remained unknown for a bit longer.[1]

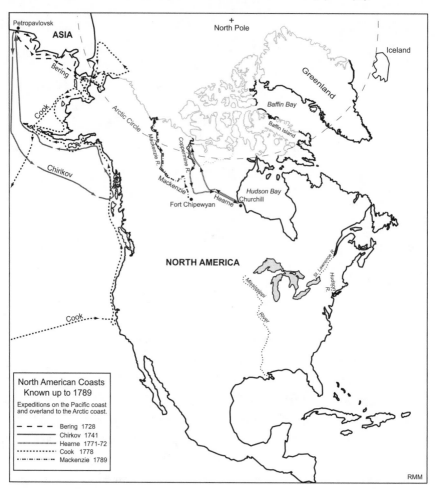

FIGURE 9.1 The state of North American coastal mapping up to 1789, showing routes of Bering, Chirikov, Hearne, Mackenzie, and Cook.

Thirteen years later, Bering undertook another voyage for the tsar with two ships, one under the command of Aleksei Chirikov (1703–1748). The boats became separated, resulting in two separate sightings of North America. Chirikov sighted the coast of southern Alaska at 55° 21′, but failed to make landfall. In the effort to reach the Alaskan shore, both his ships' landing boats wrecked and lost their crews. Chirikov continued sailing northwest along the coast and southwest along the Aleutian Islands, resulting in a rudimentary but reasonably accurate map of the south Alaskan coast. An enlarged map of their voyages is shown in figure 9.2.

Bering made only one landfall to resupply the ship's water and quickly continued. His crew suffered from scurvy, and the undermanned ship was greatly handicapped. Bering tried to make it back to port but had the misfortune to wreck on

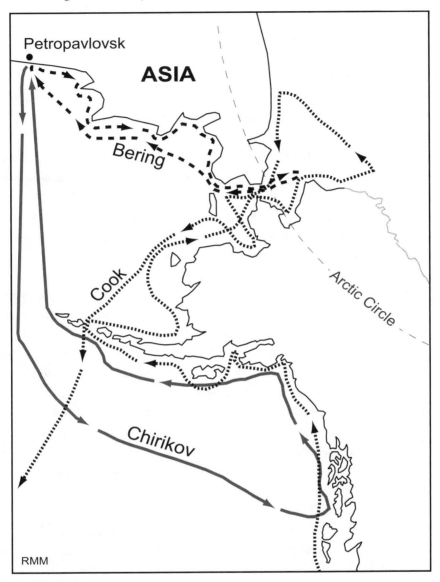

FIGURE 9.2 An enlarged map of Cook's route along the Alaskan coast suggests his intent to make a thorough map of the coastline.

an island near Kamchatka, where Bering and much of his crew died. Survivors eventually made it back to port and reported the abundance of fur-bearing seals, foxes, and sea otters, causing a surge of Russian trading ships returning to Alaska over the next two decades. Bering's voyage added little to the map, however, Bering's name is the one that continues to identify several land features: a strait, a sea, a glacier, and a lake. A small island south of the Alaskan Peninsula is named for Chirikov.

Bering and Chirikov established a few coordinates marking the western limit of the North American continent, but the northern limit remained a mystery. Samuel Hearne's expedition in 1770 provided a clue to the mystery by establishing one point of reference along a vast, unknown northern coast.

Samuel Hearne (1745–1792) learned navigation with the Royal Navy, which he entered at age 11. Later he left the navy and joined the Hudson Bay Company as a mate aboard one of their ships. He attracted the attention of officials in the Company, and they selected Hearne to investigate reports of copper deposits in the Canadian far north. Hearne proved to be an able explorer, observer of natural environment, and leader of men.

The party had an arduous journey, beginning 1770, with poor provisions and shelter through terrain that offered little sustenance. Upon reaching a river flowing northward through the flat and treeless tundra, Hearne was disappointed to find only a small amount of native copper. He nevertheless named it the Coppermine River and continued downstream to its mouth. Although Hearne made no survey along the coast, this one coordinate was significant by establishing, for the first time, a mapped point marking the northern limit of the North American mainland. He returned to Prince of Wales Fort in 1772 after a journey of eighteen months in the Arctic wilderness. When John Franklin came to this spot in 1819, he found that Hearne had erred by making the coast 200 miles too far north.

Besides mapping a point on the northern coast of North America, Hearne ascertained that no navigable water passage existed between Hudson Bay and the Arctic Ocean. These achievements helped provide a glimpse of the enormous extent of the North American continent. He also disproved the existence of the supposed Straits of Anian that fifteenth-century cartographers had created.

A later river expedition by a Scotsman named Alexander Mackenzie (1764–1820) again went through the continental interior to provide a second point on the Arctic coast. Mackenzie, an employee of the Northwest Fur Company, departed in 1789 from Fort Chipewyan on Lake Athabaska, accompanied by twelve Indians. They traveled in canoes on the Slave River to the Great Slave Lake, then continued northwestward along a river (Mackenzie River) that the Indians said entered the far northern ocean. Mackenzie held high expectations that the river would actually enter the Pacific Ocean at Cook Inlet, Alaska. He named it Disappointment River when he discovered that it emptied into the Arctic Ocean. It was later renamed Mackenzie River in his honor.

As the expedition progressed northward toward the lands inhabited by Eskimos, the Indians balked at continuing because of their extreme hostility toward the Eskimos. Samuel Hearne had also experienced the severe enmity between these two native groups when the Indians with him massacred a group of Eskimos.[2] Mackenzie was able to persuade the Indians to continue, and eventually they reached the coastal plain and the broad delta where the river reaches the Arctic Ocean. The Eskimos in the area told Mackenzie of other white men who had arrived in big boats some eight or ten years earlier.[3] Probably those men were Russian traders or fishermen probing eastward from their trading posts in Alaska.

The main mapping achievement of the journey was to establish a second set of coordinates marking a more westerly location along the Arctic coast of North America. Taken together, Hearne and Mackenzie demonstrated conclusively that no water passage connected the Atlantic and Pacific Oceans through the continent.

10

James Cook Maps a Huge Swath of the Northwest Coast, 1778

JAMES COOK (1728–1779) is an imposing figure in the naval history of England because of his three grand Pacific expeditions, the third making a western probe into the Arctic. His greatest legacy is the excellent maps he made wherever he went, some of them with meticulous detail. The earliest of this high-quality mapping came at a time before the marine chronometer was available to calculate longitude. Cook was an explorer at heart. He once expressed an ambition to go not only farther than any man had gone before, but also as far as it was possible for man to go. His goal was to map the Pacific as accurately as possible. Other mariners' maps were so unreliable that some islands had never been mapped, and other mapped islands (for example, the Solomon Islands) were plotted so poorly that they had to be rediscovered two centuries later.

In 1745, at the age of 17, young James Cook lived in the small village of Staithes on the Yorkshire coast near the whaling port of Whitby. After a year and a half in that village, and living in the continual presence of seamen, Cook signed as an apprentice with shipowner John Walker. Walker sent ships to countries around the North Sea trading in coal. Cook took an immediate interest in learning navigation techniques, and in 1755 he was offered command of one of Walker's ships but declined in favor of joining the Royal Navy as an able seaman.[1]

Cook was clearly a rising star in the Royal Navy. Within a month, he had learned his job well enough to be upgraded to master's mate, a considerable step up in responsibility. Two years later, he passed the examination for ship's master with responsibility for navigating and sailing a ship. In 1758, on one of his missions in the Seven Year's War, Cook met a surveyor who was making a map of the fallen French fort at Louisbourg, Nova Scotia, and there he undertook to learn mapping skills. Cook learned plane table mapping techniques. He also studied trigonometry and astronomy on his own time. Therefore, he could sail a ship expertly and map a coast line when most captains left map making to others. From that time, Cook maintained an ardent interest in hydrographic mapping and soon had completed a detailed map of the Bay of Gaspé. Over the next years, he made detailed maps of parts of the Gulf of St. Lawrence, the St. Lawrence River mouth, and the island of Newfoundland. These maps were remarkable for their time, and they established Cook as a mariner of great skill and intelligence.[2]

The Newfoundland map, which Cook worked on for nearly four years beginning in 1763, bears special mention because of the meticulous care taken to make it a detailed map showing every bay, promontory, and cove. Cook used the triangulation method for mapping Newfoundland and spent much of the time on shore measuring baselines and marking prominent features for sighting. At the time of his Newfoundland survey, Cook had no certain means of finding longitude until a solar eclipse occurred in the summer of 1766, allowing computation of the island. Triangulation surveys are extremely time consuming and require the support of a number of people in the field to make it work.[3] Although Cook took great pains for accuracy during most of his surveys at sea, he normally used running surveys, sacrificing detail for the sake of saving time. A running survey during a voyage is the only feasible option for mapping immense lengths of coastline in a reasonable time.

Beginning in 1768, Cook led three expeditions that became models of hydrographic mapping. His ambitious objective was to survey the vast Pacific Ocean, search for an expected, but undiscovered, southern continent (Antarctica), and make another attempt to find the Northwest Passage.[4]

His three expeditions explored and mapped much of the Pacific, including Australia, New Zealand, and Hawaii. Most important to the objective of this book is his third voyage, in which he surveyed the North American coast from Cape Blanco, in present day Oregon, through Bering Strait to Icy Cape on the north coast of Alaska. Cook closed a great gap in the North American map comparable to the great gap filled by Verrazzano two hundred years earlier. His terminal point in the Arctic, Icy Cape, long stood as a benchmark from which other mappers tried to extend maps eastward. Another significant aspect of this voyage was that Cook was carrying, as he had on the second voyage, the K1 (K for Kendall, the maker) replica of the John Harrison H4 clock for calculating longitude.[5] Cook's endorsement of the new clock opened the way for reluctant mariners to adopt this innovative method for measuring longitude.

The third voyage began in Plymouth, England, on July 12, 1776, with the announced intention of returning a native Tahitian, Omai, to his homeland. Omai had been brought by Cook to England on a previous voyage and had spent two years there as something of a celebrity. The unannounced plan was to sail two ships north along the west coast of North America in search of a western entrance to a Northwest Passage. Cook commanded the *Resolution* and Captain Charles Clerke commanded the *Discovery*. The expedition left Omai in Tahiti, turned north, and in January 1778 became the first Europeans to visit the islands that Cook named the Sandwich Islands in honor of the Earl of Sandwich, First Lord of the Admiralty. The first landfall was at Waimia Harbor in Kauai. An excellent record of this voyage was kept by midshipman George Gilbert, a young officer-in-training aboard the *Resolution*.

After several friendly contacts with the native Hawaiians and resupplying wood and water, Cook's expedition continued toward North America. He began his traverse up the coast at Cape Blanco, the northernmost point reached by Spanish mariners and by Sir Francis Drake in the sixteenth century. Cook's group were the first Europeans to enter Nootka Sound, which Cook named King George Sound, on the west coast of Vancouver Island. They spent five weeks in Nootka Sound cutting and shaping trees to replace damaged masts and spars. Cook and his officers set up an observatory on shore and made repeated astronomical observations. During the time in Nootka Sound, Cook and his men had friendly relations trading with the Indians. One apparent threat arose when an unfamiliar party of Indians arrived in canoes, and the ship's company, because of the hostile appearance of the Indians, thought they might be combining forces for an attack. The crew took care to arm themselves and set a contingent around the observatory on shore, but were under orders from Cook to hold their fire. As it turned out, the two groups of Indians had a dispute only between themselves over trading rights with the ships. Gilbert observed that the Indians tried to exert their will against each other by pulling hair until one side quit. He had seen other Indians settle minor disputes by boxing.

In April, the ships again set sail and continued up the coast. Gilbert described forests so thick that they could not walk more than a hundred yards from the narrow and infrequent beaches. The coastal area was mountainous, and the higher elevations had snow. He told of finding wild strawberries, gooseberries, and currants. He saw bears, wolves, deer, foxes, and beaver and purchased pelts of these animals from Indians along the way. The crewmen immediately used the furs to make warm clothing against the increasing chill as they headed north. Gilbert gave detailed descriptions of the Indians he saw. In most cases, he compared them with natives of other places he had visited. They were fewer in number than the populations of the tropical islands. Their hair was matted with a red mixture much like that used by

New Zealanders both in color and smell. They wore cloaks made of tied and woven grasses like the New Zealanders. They wore round caps with a point at the top like the Chinese. The women were dirtier and "far unlike the blooming beauties of the tropicks." He compared their broad flat-bottomed canoes to a Norwegian "yaul" cut from one tree. Gilbert described the language of the coastal Indians as harsh and guttural sounding to his ear, with words that even some of the Indians had trouble pronouncing. Gilbert gave special attention to the Indian's weapons. Their bows, the best Gilbert had ever seen, were made very strong and flexible by having whale sinews stretched along a groove on the back of the bow. The arrows were made of fir wood and headed with a hard wood, bone, or flint. Their spears were pointed in the same way. He went on to describe in detail the Indians' way of fishing. Such descriptions of the environment and the people became valuable additions to the discoveries and provided much more than simply a map of the explorers' traverse. Gilbert indicated his awareness that Bering and Chirikov had each reached some of these shores on Russian expeditions in 1728 and 1741.

A severe storm forced both of the ships to sail far away from the shore and therefore leave several hundred miles of the shoreline unmapped before they could again resume their coastal survey. The storm caused serious leaks in the *Resolution*, flooding part of the ship and keeping the crew manning pumps and buckets for two days.

Finally reaching lands covered completely with snow at 60° North latitude, they saw an opening of "very promising appearance" that raised hopes of finding the long-sought passage. The "passage" was the large inlet now called Prince William Sound. Fog prevented them from seeing far, but the next morning, clear skies allowed them to see that the sound was no passage. Later, an even more promising opening appeared, Cook Inlet, which the ships traversed, and again met with disappointment. At the head of Cook Inlet, a group of men from the ships landed and took possession of the country in the name of the King of England. Gilbert mentioned that a few local natives watched the proceedings, but showed no comprehension. Taking possession at this time was questionable, considering that Russia was already actively trading in the area and believed the area to be in their possession. In this area, the expedition encountered Eskimos, rather than Indians, for the first time. Gilbert observed some differences in their appearance and demeanor, but gave more attention to the difference in their boats—kayaks, which were made of hides and carried either one or two people, and umiaks, which could carry fifteen or twenty men. A detail of Cook's route along the Alaskan coast can be seen in figure 9.2.

Cook proceeded past the Shumagin Islands, so named by Vitus Bering for one of his sailors who was buried there in 1741. Cook continued around the southwestern tip of Alaska and turned northward. Again Gilbert told of meeting Eskimos (Aleuts) and how much more civil they were than Indians they had met on the coast farther south. He described their houses as mounds of earth covered with turf. The houses varied in shape from circular to oblong or square and were about 25 feet long with a floor 6 feet below ground level. Apparently Gilbert went inside some of the houses and found three or four families separated by partitions, like stalls in a stable. Beds consisted of raised earth ledges covered with skins.

Continuing northeast along the north shore of the Alaska Peninsula, Cook followed the coast to the place Gilbert identified as the most western extremity of North America—named Cape Prince of Wales by Cook. The expedition continued northward, but had periods during which they could not see the shore because of the fog. On one occasion, they came within minutes of running aground in the fog. At latitude 70° 40′ N, they met continuous ice locked to the shore and extending out to sea as far as they could see. They had reached their limit at a point Cook named Icy Cape for obvious reasons. The shore was sufficiently bound in ice that it was impossible to reach it by boat. It was then August 18, 1778, and time to leave the extreme parts of the Arctic.

While in the area, Gilbert went to some length describing walruses, which they called sea horses. They killed several for the oil and meat, but Gilbert found the meat had a disagreeable taste and odor. On the southward part of the voyage, Cook revisited some parts of the coast that they had missed because of fog, and entered Norton Sound where they found wood for the first time since leaving Cook Inlet. They resupplied wood and water and found a rich supply of berries which gave them all great pleasure. As they reached the Unalaska area in the Aleutians, the natives there brought the captain a cake made of rye flour. Cook immediately assumed that it had been given by Russians on the island and asked the Aleuts to lead the way. Cook sent one of the marines who returned in two days with three Russians. Neither party could speak the other's language, so communication was done by signs and maps, which everyone understood very well. Cook learned that a group of about fifteen Russians lived on Unalaska and traded along the coast.

Cook finally left Alaskan waters for good on October 31, 1778, heading back to Hawaii for the winter. In this third voyage, Cook had charted over 4,000 miles of the North American northwest coastline, determined the extent of Alaska up to Icy Cape, and closed the gap from the Russian survey by Chirikov and Bering to the Spanish survey at Cape Blanco, Oregon. They may have felt disappointed at not finding a passage through the continent, but their contribution was immense.

The return to Hawaii in February 1779 had an disastrous outcome. The Hawaiians seemed to be less welcoming than on the first visit, and disagreements and minor arguments with crewmen sprang up. The Hawaiians captured a cutter, the small boat carried on deck, and resisted when the ship's company demanded it be returned.

Cook took a contingent of armed marines ashore to assert his will and regain the boat. As sometimes happens in a tense situation, Cook unwisely fired a shot and marines opened fire. Numbers favored the Hawaiians. The marines and Cook began to retreat, but Cook was stabbed and killed before he could return to the boat. Gilbert's account said that Cook fell into water near the shore and the Hawaiians held him under to be sure he had died. Captain Clerke, now commander of the expedition, later retrieved parts of Cook's body and held an elaborate burial ceremony in Kealakekua Bay on the island of Hawaii.

PART IV

The British Surge to Find the Northwest Passage also Makes Maps, 1818–1845

Watching a coast as it slips by the ship is like thinking about an enigma. There it is before you—smiling, frowning, inviting, grand, mean, insipid, or savage, and always mute with an air of whispering, "come and find out."

—JOSEPH CONRAD

11

John Ross Sees a Mirage, 1818; John Franklin Makes His First

Expedition, 1819

A LONG LULL in Arctic exploration ended when England again took interest in exploring and mapping the islands of the North American Arctic. Thirty years had elapsed since any previous effort to find the passage. For much of this time, England was involved with the Napoleonic Wars—not to mention another conflict with the United States—and the Royal Navy was focused on warships. Finally, after the Trafalgar sea victory and Napoleon's final defeat in 1815, the navy was again faced with a surplus of men and ships. The navy needed to keep ships active and men trained, and the peace time remedy was exploration. The first effort in 1818 was to try crossing over the pole, as Hudson had done two hundred years earlier but to no avail. Little of the Canadian Archipelago had been mapped before the major push by the Admiralty beginning in 1818. Sixty years later they had mapped about 90 percent of the coastlines in the Canadian Archipelago as shown in figure 11.1.

In 1818, Captain John Ross (1777–1856) commanded the former whaler, *Isabella*, on an expedition, along with a second ship, *Alexander*, commanded by Lieutenant William Edward Parry. Together they planned a search for the Northwest Passage by sailing westward beyond Davis Strait. James Clark Ross, John's nephew, sailed with him on the *Isabella*. Although they mapped several new islands and corrected some errors on earlier maps, Ross missed the prize by turning back too soon. As the two ships sailed about 50 miles into Lancaster Sound, Ross saw what he thought was a range of mountains blocking the way ahead. He recorded the perceived range as the Croker Mountains, which in fact may have been a bank of ice fog or an optical

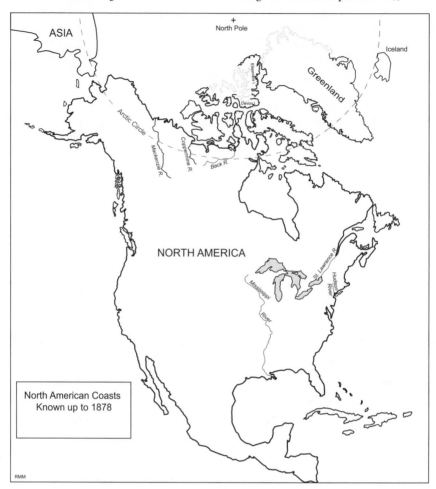

FIGURE 11.1 After the intensive exploration and mapping effort of the nineteenth century, North America had only a few unmapped coastlines in the farthest reaches of the Arctic islands.

illusion. This mistake ruined Ross's credibility as a seaman, and he returned to England having missed the real entry to the Northwest Passage.[1]

William Edward Parry, captain of the *Alexander*, convinced the Admiralty that Ross had made a serious misjudgment, and that the Croker Mountains did not exist. Parry insisted that Ross had turned back just as the expedition was on the verge of a major discovery. Further, Parry said that Ross had been the only one on the expedition who believed the Croker Mountains existed. On that basis, the Admiralty bypassed Ross and sent Parry on a second expedition in 1819 to explore Lancaster Sound farther westward. Parry had exceptional good luck by arriving during a year in which the ice had receded earlier than usual, and he made passage westward to Melville Island, almost half the distance through the Passage to the Bering Strait.

This proved that Ross had returned too hastily. Parry returned to high praise and a monetary prize for him and his crew. The unfortunate Ross was still in a state of disgrace over his failure.

JOHN FRANKLIN'S FIRST OVERLAND EXPEDITION

For forty years (1804–1844), Sir John Barrow held the post of Second Secretary to the Admiralty, and during that time he promoted the Royal Navy's role in the exploration and mapping of the Canadian Arctic, particularly a sea passage through the Arctic. His approach included expeditions from the Atlantic, the Pacific, and overland from Hudson Bay along rivers to the north coast of the continent. Alexander Mackenzie had established latitude and longitude at the mouth of the Mackenzie River in 1789, and Samuel Hearne had determined coordinates at the mouth of the Coppermine River in 1769. Sir John Barrow had the vision to see the importance of expanding on those two points and mapping a northern coastline for North America.

In 1819, John Franklin (1786–1847) was selected to command a land expedition to find and map a portion of the uncharted north coast. His instruction: determine the latitude and longitude of the north coast of North America and the trending of the coast east and west. Also, he was directed to erect cairns containing his findings for Lieutenant Parry, whose 1819 expedition planned to reach the coast from the sea. Franklin's group was to record observations of temperature, wind and weather, the aurora borealis, and geomagnetic variations. They were to make drawings of the terrain, the natives, and other things of interest along the way. On the Coppermine River they were to look for the source of copper the Indians had brought to Hudson Bay for trading. Their route of travel is shown in figure 11.2.

In preparation for the expedition, Franklin sought out Alexander Mackenzie, who was then nearly 60 years old, for advice and firsthand information on overland travel in the Arctic. In addition, the staff of the Hudson Bay Company was to give Franklin any necessary assistance and advice regarding a route to the Arctic shore and provide an escort of Indians as guides, interpreters, and game hunters. The expedition was to begin from York Factory on Hudson Bay.

At this time, the Hudson Bay Company and the North West Company were both operating in the region and in fierce competition. They had trading posts adjacent to each other in many parts of western Canada, and occasionally armed skirmishes broke out between them. Unfortunately for Franklin, the conflict between the two companies was at its height during this expedition in 1819–1821.

They planned to travel by boat as much as possible up the rivers, starting at Hudson Bay and moving to established interior outposts already built by either Hudson

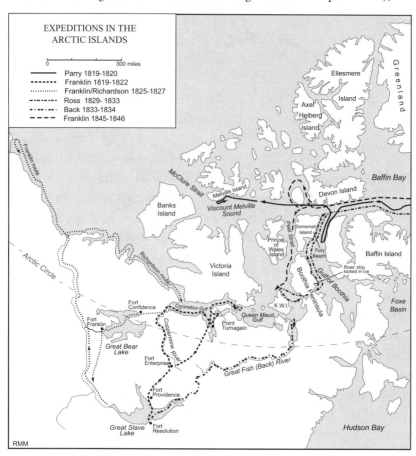

FIGURE 11.2 The Admiralty's first major penetration into the Arctic archipelago in 1818 was commanded by William E. Parry. He almost made it through the Northwest Passage. Subsequent voyages by Parry and others had less luck. Others in the first wave of that period were each by George Back, John Ross, and three by John Franklin.

Bay Company or the North West Company. Franklin was assured by the North West Company at York Factory that they would help the expedition and advise on routes and conditions in the interior. Franklin issued a directive to his officers that they should in no way interfere or participate in the quarrel between the two companies. The governor of the Hudson Bay Company had assured Franklin that certain provisions would be available for purchase at interior posts.

Franklin hastily left England in May 1819 after only three months' planning, and by early September he departed from York Factory on the long trek inland. The party consisted of five naval men and fifteen French Canadian voyageurs from the North West Company. They went up the Hayes River to Norway House trading post on the northeast corner of Lake Winnipeg; to Cumberland House on the Saskatchewan River; to Fort Chipewyan on Lake Athabasca; to Fort Providence on the

Great Slave Lake, then up the Yellowknife and Coppermine Rivers. While going upstream, a midshipman in charge of the survey took compass bearings at every turn of the river and gauged the distance traveled in each direction. These data, combined with regular determinations of latitude and longitude by means of sextant and chronometer, gave them information for charting their course.

At this stage of the journey, Franklin wrote of his impressions of the beautiful landscape about 250 miles inland from York Factory.

> Steel River presents much beautiful scenery; it winds through a narrow, but beautiful well wooded valley, which at every turn disclosed to us an agreeable variety of prospect, rendered more picturesque by the effect of the season on the foliage, now ready to drop from the trees. The light yellow of the fading poplars formed a fine contrast to the dark evergreen of the spruce, whilst the willows of an intermediate hue served to shade the two principal masses of colour into each other. The scene was occasionally enlivened by the bright purple tints of the dogwood, blended with the browner shades of the dwarf birch and frequently intermixed with the gay yellow flowers of the stubby cinquefoil. With all these charms, the scene appeared desolate from the want of the human species. The stillness was so great that even the twittering of the whiskey-johneesh or cinereous crow caused us a start.[2]

Because the rivers were running low at this late season of the year, heavily loaded boats frequently ran aground, so the crew unloaded and cached some provisions. This was done in anticipation of replacing them, as promised by the Hudson Bay Company. By October 23, 1819, they reached Cumberland House, 700 miles from York Factory. By this time the rivers were frozen and travel by water was impossible, making it prudent to winter at Cumberland House. In January, not the best time for travel, Franklin trekked ahead with four men to Fort Chipewyan, about 500 miles as the crow flies, to arrange for Indian guides for the next season.

The following spring, the rest of the party reached Fort Chipewyan, on Lake Athabasca, and proceeded on to Fort Providence, owned by the North West Company, at the west end of Great Slave Lake. Reaching this point late in July, they moved on to Fort Providence. On August 20 they built a log building on the Winter River (naming it Fort Enterprise), as it was too late in the season to reach the mouth of the Coppermine River. The reduced provisions forced them to ration food, and by the end of winter they were in desperately short supply. So far, the provisions promised by the North West Company had not been delivered.

They left Fort Enterprise in June 1821, heading toward the mouth of the Coppermine River, and sent a courier with the message that provisions should be brought to Fort Enterprise by September as a food supply for the next winter. A clerk of the North West Company was to forward the expedition's completed journals and charts to London.

Along the Coppermine River they mapped its course and sketched the terrain around it. At the mouth of the river, over 334 miles from Fort Enterprise, Franklin wrote that "the canoes and baggage were dragged over snow and ice for one hundred and seventeen miles of this distance."[3] They measured latitude at the river's mouth and found the coast to be considerably farther south than Samuel Hearne had measured eighty-two years earlier. Franklin named a prominent nearby cape for Hearne.

Now they could use the canoes to make a running survey along the coast line from the mouth of the river eastward, along the coast of Coronation Gulf—a much easier task for the sailors than overland travel. Franklin wrote that they embarked on the "Hyperborean Sea"[4] (extreme north sea) and were pleased to return to their mariner's element. But several hundred miles of sea travel amid floating ice, gale winds, waves, and surf became more hazardous in canoes. When the party ended the survey at a place Franklin named Point Turnagain, he apparently was concerned that it might appear he had not gone far enough. He wrote, "Though it will appear from the chart that the position of Point Turnagain is only six degrees and a half to the east of the mouth of the Coppermine River, we sailed, in tracing the deeply indented coast, five hundred and fifty-five geographic miles."[5] Later in his narrative, Franklin wrote, "I trust it will be judged that we prosecuted the enterprise as far as was prudent, and abandoned it only under a well founded conviction that a further advance would endanger the lives of the whole party, and prevent the knowledge of what had been done from reaching England."[6]

At the end of August 1821, they began the return to Fort Enterprise by an overland route of more than 200 miles with no river to follow—over barren, treeless tundra without knowledge of the terrain ahead—in what proved to be a disastrous journey. They had intended to return along the river as they had come, but their scanty provisions prompted them to take a more direct route, with hopes of finding game along the way. The party was nearly out of food and the winter gales had begun in early September. They were reduced to eating berries and lichen with an occasional deer to revive their energy, and occasionally resorted to eating rotting deer carcasses found along the way. Eventually they were eating leather from buffalo robes and boots. Because there was almost no wood on the tundra for fuel, most of what was eaten was uncooked. Franklin had a narrow escape when he fell into a torrential stream. He got out safely, but lost the case carrying his journal. At another time, he had several fainting spells from exhaustion.

Franklin sent midshipman George Back ahead to find the Indians that had agreed to help provide some food. On October 6, Franklin reported that the party ate what was left of their old shoes (meaning moccasins of untanned leather) and any other leather they could find. They were now within a few days of Fort Enterprise, where they expected to find a cache of food carried in over the summer by agents of the North West Company.

Several men stopped to rest because they were too weak to go on. Only Franklin, four voyageurs, and some mariners reached Fort Enterprise. Their intention was to take food and supplies to the men left behind. To their horror they found that no provisions had been brought to Fort Enterprise. All they found were some deerskins, which they roasted over a fire made from floorboards of the log house. For the next weeks they were kept alive on a diet of deer hide, lichen, and an occasional partridge bagged on their hunts.

In the meantime, at the camp of those left behind, arguments and shootings resulted in two men, a midshipman and a voyageur, being killed. The remainder from that group arrived at Fort Enterprise only to find Franklin and the others in a sad state of emaciation.

In early November, three Indians, sent by George Back, finally arrived with some meat. The starved group wolfed down the meat while the Indians built a fire and began caring for the starving men as though they were children. Franklin wrote that "the Indians treated us with the utmost tenderness, gave us their snowshoes, and walked without themselves, keeping by our sides that they might lift us if we fell."[7] One of the Indians returned to George Back with a note requesting more food. Of the original party of twenty, eleven had died, including the midshipman killed in the shooting, and ten voyageurs. Four out of five naval personnel survived. The fact that none of the naval personnel died of starvation was later attributed to the good discipline of the navy. Three of the voyageurs had departed for Montreal separately. In July 1822, Franklin and five others arrived safely in York Factory after having traveled 5,550 miles over land, river, and sea to map 350 miles of the north coast of America.

George Back's courage, determination, and ultimate success in finding the Indians and procuring a food supply was doubtless responsible for the party's survival. During his return to Fort Providence, Back was himself in danger of starvation. He traveled 1,200 miles on snowshoes in temperatures below -50°F, with no shelter but a blanket and a deerskin, and faced two or three days at a time without food. He succeeded just in time to save the lives of Franklin and the rest of the crew at Fort Enterprise. On returning to England, midshipman George Back received a well-deserved promotion to lieutenant. Franklin was a hero for having saved his crewmen from starvation and received the moniker, "the man who ate his boots."

12

William E. Parry Has Beginner's Luck, 1819, 1821, 1824

William E. Parry (1790–1855) went to sea at the age of 13 and reached the rank of lieutenant by age 17. At age 25 he was given command of the *Alexander* for the 1818 expedition led by John Ross. His success on that voyage and his belief that Ross had failed to discover the Northwest Passage by turning back too soon brought him the command of a return expedition in 1819 to probe the route Ross had presumed was blocked by mountains. Parry's objective was to discover the Passage by sailing two ships, *Griper* and *Hecla*, through Lancaster Sound, find the north coast of North America, and go to the mouth of the Mackenzie River. The intent was to reach the coastal sites established by Mackenzie's expedition of 1789 and John Franklin's current expedition to the mouth of the Coppermine River. Their routes are shown in figure 11.2.

His expedition enjoyed uncommonly good fortune. The year 1819 was an unusually good year for Arctic sailing. Parry acknowledged that the seas had less ice than on his previous voyage and that the weather was better than usual. He passed through Davis Strait and Baffin Bay unimpeded and once in Lancaster Sound sailed through the supposed location of the "Croker Mountains" that Ross had mistakenly mapped. They found continuous open water beyond 110° west longitude, farther west than anyone had ever sailed from the east in the Arctic. Achieving this distance qualified Parry and the crew for a monetary reward by meeting a preliminary objective set by the Admiralty.

They spent the winter at Melville Island. The ships were moved into a quiet bay, frozen into the ice, the hulls covered with padded cloth and banked with snow for insulation and protection from the wind. During the nearly ten months locked in ice, their time was spent schooling the crew, putting on plays and skits, playing musical entertainment, and making scientific observations of magnetism, atmospheric temperatures, and air pressure. Hunting caribou and musk ox provided fresh meat.

Ships wintering in the Arctic might be frozen in place by late September and not sail again until the end of July. Despite the schooling and other onboard diversions, ten months of confinement and harsh weather wore heavily on the crew. Parry compared winter in the Arctic to the stillness of death, with a near total absence of any movement. Occasionally they would see Arctic foxes or wolves, but most of the time nothing moved. Parry took pride in the good spirits and good health maintained by all the crew during those long months. It was remarkable that only one man was ill throughout an eighteen-month expedition.

They learned new ways to cope with the cold. Ordinary leather shoes were inadequate against frostbite. They instead created canvas boots with rawhide soles and blanket lining, similar to snow boots used today. They also used wolfskin blankets for bedding to stay warmer at night and devised ways of reducing moisture accumulation in the living quarters by circulating vapor from washing, cooking, and respiration past cold surfaces where it condensed and turned to ice. They used kiln-dried flour stored in air tight tins for baking bread in place of sea biscuits.

Canning meat and vegetables, invented in 1810, made it possible to winter successfully in the Arctic. Without canned food supplies, previous Arctic expeditions had to return home before winter or face starvation and scurvy. Voyages to temperate climates and tropics could replenish fresh food along the way and thus stay for longer voyages. The first preservation of food by canning resulted from a 12,000 franc prize offered by Napoleon for the discovery of a way to preserve an adequate supply of food for a large army on the move. The Frenchman Nicolas Appert made a breakthrough and won the prize by observing that food cooked in a sealed jar or canister did not spoil. The idea spread to Britain, and in 1811 the business of Donkin, Hall, and Gamble began canning food under contract with the Royal Navy.

Another asset for long voyages was the discovery that lemon juice was an antiscorbutic. Various cures for scurvy were known in Europe by the mid-eighteenth century when voyagers to East Asia saw antiscorbutics in use. Another cure available to northern explorers was scurvy grass, which grew on the Arctic tundra.

By August 1, after the ice thinned enough to saw a path out of the bay, the *Griper* and *Hecla* were again in open water. Once underway, they hoped to continue westward through the Passage, but were confronted by permanent pack ice 40 to 50 feet thick. They had no option but to turn back eastward to Lancaster Sound and home. On the journey home they continued mapping and naming coasts, capes, bays, and islands in the Canadian Archipelago.

Parry's first expedition achieved much and was noteworthy for factors other than the good health of the crew and successful wintering in the Arctic. He made an enormous leap in the knowledge and mapping of the North American Arctic. Before this voyage, the land and sea beyond the east end of Lancaster Sound was unknown. After Parry's voyage, the map west of Lancaster Sound showed the coast lines facing Lancaster Sound and Melville Strait, but gave no indication if they were islands or parts of the mainland. His effort was only the beginning of a major expansion of mapping in the Arctic.

PARRY'S SECOND VOYAGE, 1821–1823

After Parry's stunning success the previous year, the Admiralty pushed forward with a second expedition for Parry, who by now had star status in the Royal Navy's eyes. The solid wall of pack ice that Parry found beyond Melville Island convinced the Admiralty that a route farther south would possibility be ice-free. In 1621, soon after Henry Hudson's discovery of Hudson Bay, Captain Luke Foxe probed the northern waters of the Bay into the area now called Foxe Channel, but no one had gone farther. Parry's instructions directed him to push farther into the Foxe Basin, searching for an outlet to the west.

Parry was given command of the expedition with two ships, *Fury*, commanded by H. P. Hoppner, and *Hecla*, commanded by G. F. Lyon. These ships had formerly been bomb ships, meaning they had carried large mortars for firing on shore defenses, rather than cannons for firing on other ships. Parry insisted that the two ships be identical in size and rigging, so one ship could accommodate both crews and equipment if the other ship should have to be abandoned. They rigged them with fewer sails so fewer men would be needed to man the sails, also requiring fewer provisions. Six inches of extra planking reinforced the sides of the ships against the pressures of ice. Rudder supports were made larger for easy removal of the rudder before becoming locked in the ice. Each ship was outfitted with the same sails and masts so all items would be interchangeable. If one ship were crushed by the ice, everything could be transferred and used on the remaining ship. Each was independently provisioned in case they should become separated. The design of coal stoves was improved to aid the circulation of air throughout the living quarters, and side walls were lined with cork to insulate against the cold.

The ships carried enough canned meat for each man to have two pounds of meat (mostly salt pork) per week, and vegetable and meat soup sufficient for three years. They continued carrying kiln-dried and sealed flour for making the bread. Provisions also included fresh beets and potatoes to be used in the first three months. Lemon juice in Parry's first voyage had been carried in glass jars, and many of them froze and broke. For this trip, lemon juice was stored in wooden kegs with rum added to prevent freezing—probably the first use of a rum collins. They also carried antiscorbutic vegetables, plus mustard seed and cress seed from which antiscorbutic greens could be grown on board, warmed by the heating ducts. By the following spring, the seeds had produced one hundred pounds of fresh antiscorbutic food. Both ships were fitted with snow melting tanks around the stoves' chimneys, capable of producing sixty-five gallons of water per day. A supply ship, *Nautilus*, accompanied them, carrying much of the provisions, a supply of coal, plus two live steers. When they encountered the Arctic ice, the supply ship transferred its load to the others and returned to England. Another supply ship planned to meet the expedition at Bering Strait, the west end of the Passage. In many ways, this expedition was much better prepared than previous ventures.

Parry was well aware that his first voyage was highly successful that particular year due to chance openings in the ice. He also acknowledged that this voyage could only lead to disappointment unless he sailed all the way through the Passage.

Parry's instruction from the Admiralty was to pursue two objectives: find the Passage at the north end of Hudson Bay and map the northern edge of North America. Scientific observations should be made, but not at the expense of the first two objectives. Push on to the Passage! His instructions also directed him to erect flag poles when he reached the north coast, with messages in buried bottles for John Franklin to find. Franklin was making his first overland expedition at the same time, and it was hoped that the two expeditions could join their maps of the coast.

While passing through Hudson Strait, Parry wrote about adjusting to the bleakness of the Arctic landscape, "It requires a few days to be passed amidst scenes of this nature, to erase, in a certain degree, the impressions left by more animated landscapes, and not until then does the eye become familiarized, and the mind reconciled, to the prospects of utter barrenness and desolation, such as these rugged shores present"[1]

By the first week of October, the expedition had surveyed about 600 miles of uncharted coastline. Parry felt they had made a satisfactory beginning. Now they stopped for the winter near a small island, called Winter Island, in the Foxe Basin. As before, they cut a channel through newly formed ice to reach a safe harbor. The ships were readied for winter by piling snow along the sides up to the gunwales, and covering the main deck with a canvas tent. Below decks, the living quarters were warm and dry, thanks to coal stoves circulating warm air throughout. Once settled, the ships' company returned to a winter routine of entertainment with theatrical productions, musical performances, and the customary schooling for sailors who could not read. Parry felt great satisfaction that each man could read the Bible by the end of the expedition. All of these activities kept the men occupied and maintained high morale. Regarding their acting performances, one officer wrote in his diary, "The audience were ready to be amused by any novelty, and in an especial manner, to be gratified by seeing the officers, to whom they were in the habit of looking up with respect and obedience, voluntarily exerting themselves for their amusement."[2]

In January, a large group of Inuit appeared within sight of the ships with about sixty men, women, and children, plus dogs, sledges, and canoes. They built a cluster of five igloos, to the great surprise of the ship's crew who had not noticed their arrival. This provided the crew another source of diversion and some social interaction. Parry, Lyon, and several crewmen ventured out apprehensively to meet the visitors and went with them to their igloos. The Inuit came unarmed and were interested in trading. Later some of the Inuit returned with the sailors to the ships where they were shown around and entertained with music on the flute and violin. The Inuit in turn sang song after song for the crew. Parry and one of the other officers wrote the notes to the Inuit songs. This friendly beginning led to continued interaction with the Inuits, including visits to their homes. Parry attempted to learn their language and write an Inuit vocabulary useful for future explorers. When the ships' company donned their dress uniforms for Sunday services, the Inuit danced with delight at the sight of the marine's bright red coats.

Captain George F. Lyon wrote in great detail of his acquaintance with individual Inuits. In particular, he described the intelligence and musical ability of a certain woman named Iligliuk. She was able to draw, from memory, a map showing a route from their winter quarters to a northern strait that passed through to the western sea. This map caused much anticipation among the officers. They felt sure they would have great success in the coming summer in completing a traverse of the Northwest Passage. Parry could now see a real possibility of rounding the corner of the continent and progressing to the western shore of North America.

In mid-summer, the expedition sailed free of the ice and continued their exploration of northern Foxe Basin. As the ships rounded the point of land at the entrance to the strait described by Iligliuk, Parry believed he had found the northeast corner of North America and named it Cape North-East. When the expedition investigated the strait, which they named Fury and Hecla Strait after their ships, they found the detail of Iligliuk's sketched map accurate in detail, but they also found the strait to be blocked with ice and unnavigable. They stopped at the opening of the strait for four weeks chafing at the prospect of being unable to move on from the edge of the Passage.

The crews managed to tow some of the floating ice floes away from the opening of the strait by tying them to the ships and sailing eastward with a west wind. This effort opened a channel into the strait and allowed them to sail ahead under full sail. After a short distance, the ice locked shore to shore, making further progress impossible. They observed an eastward flowing current passing their ships under the ice. From this they concluded that an open sea must lie beyond the impassable strait. The only option now was to continue on foot to see the extent of this strait. Parry sent a party of men to investigate the strait by land on August 30 amid early snowfalls. They journeyed along the north shore of the strait until they reached open water, which the mapping officer labeled Polar Sea, later named the Gulf of Boothia by John Ross.

The expedition spent their second winter at the eastern end of Fury and Hecla Strait from October 1822 to August 1823. Again they had the company of Inuit, creating a pleasant diversion over the winter months. In August they made a brief and futile attempt to navigate the strait again before heading back to England. No ship managed to navigate Fury and Hecla Strait until 1948, when a United States icebreaker pushed its way through from west to east. Fury and Hecla Strait has a persistent ice floe, fed by ice from the Gulf of Boothia with the push of westerly winds and an east-flowing current. The narrow strait is about 80 miles long and averages about 10 miles wide. However, the east end of the strait is blocked by Ormonde Island, leaving a narrow opening only 1 mile wide that acts as a bottleneck, becoming choked with ice.

Parry had been hoping to enhance his fame on this voyage, but it became clear that John Franklin's harrowing, but successful, overland expedition overshadowed Parry's and made Franklin the current idol of both the Admiralty and the public. However, the next year Parry began yet another voyage in search of the Northwest Passage.

Mapping unknown coastlines in North America while en route eventually proved to be the most important contribution from the efforts to find the Northwest Passage, but each explorer felt great disappointment in failing to find it. The Admiralty had given instructions with one priority—find the Northwest Passage. The remarkable advances in mapping unknown territory and the increase in scientific knowledge of the region were not sufficient to soothe their disappointment. Their ambition to discover the Passage fueled the quest, and nothing less could satisfy. Parry was sure that one more voyage would bring success, and, like most explorers of the polar regions, he followed his second voyage of exploration with an idea about where the next voyage should go.

PARRY'S THIRD VOYAGE, 1824–1825

Parry soon proposed a third voyage via Lancaster Sound, then turning southward into Prince Regent Inlet and through the Gulf of Boothia, the water body he had seen at the west end of Fury and Hecla Strait. In his proposal he stressed his intention to choose a sailing route near coastlines. This was contrary to the thinking of the day that considered ice to be less likely in the open water. Parry's Arctic experience showed that often the only open water was near the shore. Had the wishful myth of an Arctic sea free of ice finally dissipated?

Parry reminded the Admiralty that any delay would make it more likely for another country to seize the glory of finding the Passage. Finding the Northwest Passage was no longer motivated by possible commercial advantages, but was driven purely by the prestige and recognition that would come with its discovery. The Admiralty again approved Parry's proposal and arranged for a supply ship to meet him at the west end of his voyage, Bering Strait. Furthermore, as a show of the favor Parry held with the Admiralty, he was offered the post of Hydrographer of the Navy, to be filled on his return.

Again sailing with the *Hecla* and *Fury*, this expedition proved to be Parry's great misfortune. Davis Strait and Baffin Bay had so much floating ice in the summer of 1824 that it took the ships eight weeks to cross from Greenland to the east opening of Lancaster Sound. Much of the distance required the arduous task of warping the ships through ice by securing an anchor into a large iceberg ahead and inching the ships forward though the ice with a capstan, then repeating the process. Besides the backbreaking and time-consuming labor, warping led to several injuries and damage to ships when an anchor became dislodged or a hawser broke. Parry wrote, "On one occasion three of the Hecla's seamen were knocked down as instantaneously as by a gun shot by the sudden flying out of an anchor."[3] Crews of the *Hecla* and *Fury* continued this tedious process for two months, going nearly 400 miles through Davis Strait and Baffin Bay among hundreds of icebergs, many as high as 200 feet.

This unfortunate loss of time brought them to Lancaster Sound in early September as winter storms were beginning. Storms slowed them further as they progressed through the Sound, but they finally made it to Prince Regent Inlet in October, when ice had begun to form on the sea. They pulled into a harbor for the winter and prepared the ships for the long cold months ahead. The crew began the usual routine of ship maintenance, schooling, and entertainment activities with plays and musical performances. Regular measurements of air temperature and magnetic intensity continued through the winter. Parry wrote an eloquent description of the feeling of an Arctic winter: "In the very silence there is a deadness with which a human spectator appears out of keeping. The presence of man seems an intrusion on the dreary solitude of the wintry desert, which even its native animals have for awhile forsaken."[4]

The expedition stayed at this location until July 1825, when, with great effort, they finally sawed a path for the ship out into the open water amid floating icebergs. Prior to cutting, they had spread a coat of sand to melt some of the ice and make it a bit thinner. Parry sailed the expedition south along the west shore of Prince Regent Inlet, mapping the coastline and naming bays, looking for a strait or passage to the west. On the last day of July, a storm shoved the *Hecla* toward shore where she became grounded and immovable. Later the *Fury* also became grounded and began to leak badly. High tide lifted both ships free, but damage to the *Fury* required constant use of four hand-operated pumps to keep her afloat. They had no alternative but to take the *Fury* to the nearest beach and careen her onto her side so the carpenter could repair the hull. As the ships were moving toward the shore, wind and ice violently pushed the *Fury* onto the shore. Crewmen began transferring food stores and equipment from the *Fury* to the *Hecla* to lighten her load and save the food from damage. They set up tents on the shore for additional food storage. All hands of both ships continued the exhausting work around the clock. Carpenters began work on the hull. Blacksmiths forged new bolts. Amid all this rush, a blizzard created such high surf that all work stopped. Crewmen of both ships moved into the *Hecla*, and officers moved into tents on the shore. Parry's journal described the men as too harassed and fatigued to continue work without rest. He noticed the fatigue causing a mental failure among some men who seemed dazed, failing to understand the meaning of orders.

Gale winds continued through the night, pushing ice masses into their makeshift haven. Parry consulted with Hoppner, captain of the *Fury*, and they agreed to try to move back out into open water before they were run aground. The crews wrapped the *Fury's* hull in canvas to slow the leaking, but she was driven aground despite their great efforts. As the tide began to fall, the *Fury* became firmly grounded. In the meantime, the *Hecla* made it to open water with all hands aboard and quickly became separated from the *Fury* by several miles of pack ice. Parry and Hoppner visited the *Fury* by boat and found it had moved even farther onto the beach. The situation was hopeless, and the only option was to abandon the *Fury*. The crew salvaged personal belongings and any remaining supplies, rowing out to the *Hecla* with heavy hearts and a profound sense of defeat. The crew left a large cache of provisions and supplies on the beach that became known as Fury Beach.

The entire ordeal had begun with the storm on July 31 and ended on August 21. The *Hecla*, with a double crew, turned and headed back to England. Once again, high hopes of finding the Northwest Passage had been crushed by the relentless Arctic ice. Parry's final expedition in search of the Passage had added nothing to the coastal maps nor to the scientific knowledge of the Arctic Archipelago. His dejection and resignation to fate are clear in his final statement of the account of the expedition: "To any persons qualified to judge it will be plain that an occurrence of this nature was at all times rather to be expected than otherwise, and that the only real cause for wonder has been our long exemption from such a disaster."[5]

VOYAGE OF FREDERICK BEECHEY IN THE *BLOSSOM*, 1825–1828

The Admiralty had the foresight to arrange for a resupply ship, *Blossom*, to meet Parry as he finished traversing the Passage with *Fury* and *Hecla* into Bering Strait. Had Parry actually succeeded, resupply would have been needed for the long voyage back to England. Captain Frederick Beechey commanded the *Blossom*, and he was well acquainted with Arctic sailing through his role as a lieutenant on the *Hecla* when Parry made his first voyage all the way to Melville Island.

The voyage of the *Blossom* began in May 1825, rounded Cape Horn, making a circuitous route through the Pacific, including a call at Pitcairn Island where he met the lone remaining survivor of the Bounty mutiny, along with numerous descendants. In June 1826, they stopped in the Russian port, Petropavlovsk, on the Kamchatka Peninsula. There Beechey learned that Parry had ended his expedition after the wreck of the *Fury* and had returned to England.

Beechey continued through the Bering Strait and continued as far as Icy Cape for a possible meeting with Franklin, who was engaged with his second overland expedition, but Franklin had turned back before reaching Point Barrow. Neither Franklin nor Parry made a rendezvous with the *Blossom*.

During the waiting period, Beechey sent a boat party to survey the coast eastward from Icy Cape to Point Barrow, leaving an unsurveyed gap of only 167 miles from Point Barrow to Return Reef, Franklin's western-most point from the mouth of the Mackenzie River. The survey party, led by the Master of the *Blossom*, Thomas Elson, met great difficulties with ice, and at some locations the boats had to be towed along the shore, which in some instances were cliffs. They left bottled messages in cairns for Franklin and returned to England in October 1928. After a voyage of three and a half years covering 73,000 miles, they had surveyed about 150 coastal miles from Icy Cape to Point Barrow.

13

John Franklin's Second Overland Expedition Makes a Successful

Survey, 1825

IN 1825, JOHN Franklin commanded a second overland expedition from Hudson Bay along the Mackenzie River to the north coast, where he extended his survey of the previous disastrous trip. This new expedition was described in his *Narrative of a Second Expedition to the Shores of the Polar Sea (1828)*.[1]

When Franklin heard in 1823 that the Admiralty was planning to send William Parry on another voyage in search of the Northwest Passage, he proposed that an overland expedition be sent to survey the coast between the Mackenzie and the Coppermine Rivers, and also westward from the Mackenzie. There was even the possibility that he might rendezvous with Parry, assuming Parry was successful in finding the Passage. John Franklin had to assure the Admiralty that the problems of the first overland expedition would not be repeated. Much had been learned from the first experience; the Hudson Bay Company and the North West Company were now merged, so the supply problems arising from company competition would not recur. The proposal was approved. Franklin would lead the survey from the Mackenzie westward to Icy Cape, which Captain James Cook had reached by sailing through the Bering Strait in 1778. This survey would connect all the previously surveyed parts of the north coast, leaving only the gap between Boothia Peninsula west to Coronation Gulf. Figure 11.2 in Chapter 11 shows Franklin's route of travel.

Dr. John Richardson, who had been on the first expedition, offered his services again as naturalist and surgeon. Richardson also suggested that he could conduct the survey of the eastern portion between the mouths of the Mackenzie and Coppermine

Rivers. Lieutenant George Back also applied and was selected for the second trip. That both of these well-qualified men wanted to go with Franklin again after the near disaster of the first trip is strong testimony to their confidence in Franklin. Other assignments were assistant surveyor E. N. Kendall and assistant naturalist Thomas Drummond. Peter Dease of the Hudson Bay Company was assigned to make arrangements for reliable supplies of food supplies, including fishing and hunting along the way by voyageurs (French Canadian boatmen and explorers) and Indians. The Hudson Bay Company now held full jurisdiction over the whole of central and western Canada out to the Rocky Mountains. (This immense domain of the Hudson Bay Company continued until 1870 when the Dominion of Canada was established and the Hudson Bay Charter voided.)

John Franklin contacted the Company directors in London telling of his appointment and his needs from the Hudson Bay Company. The directors then arranged with their agents in North America to establish provision caches along the expected route of the Franklin expedition. To make certain that communication was complete, Franklin himself wrote to the traders along his route outlining his objectives and hopes for assistance. This personal contact was an important component in preparing for a safer expedition. In turn, Franklin received a lot of constructive suggestions and support from the Hudson Bay traders in Canada. They agreed to plans for finding someone to fish and for a party to build the winter shelter on Great Bear Lake.

In June 1824, new boats especially designed for the rigors of coastal travel and ease of portage were sent from England by a Hudson Bay Company ship to York Factory, along with provisions and supplies. Franklin requested two large freight canoes to be available for the expedition's arrival in spring of 1825. All these preparations prior to departure contrasted dramatically from the three months of preparation for the previous expedition. Franklin designed the new boats to be better suited for the coastal surveys in the Arctic Ocean. The earlier canoes had been good for rivers, but the pounding waves and collisions with floating ice had been too much for them. The new boats, 26 and 24 feet long, were designed much like the original canoes, but were made to withstand ocean travel and to allow for easy portage. Each boat could carry an officer, crew, and two to three tons of cargo. A smaller canvas-covered boat weighing eighty-five pounds was designed for crossing hazardous rivers, and could be assembled.

Each person packed two waterproof suits made by Mackintosh of Glasgow. Food included two years' supply of wheat flour, arrowroot (a ground tuber from the West Indies popular in nineteenth-century England for biscuits, puddings, cakes, or as a thickening agent similar to corn starch), macaroni, canned soup, chocolate, essence of coffee, sugar, and tea. They also carried tobacco, wine, spirits, tents, books, writing paper, fishing gear, and items for trading (cloth, blankets, shirts, belts, combs, mirrors, beads, knives, hatchets, and hand tools such as files, saws, shovels, and flints).

Franklin's party (Lieutenant George Back, Dr. Richardson, Mr. Kendall, Mr. Drummond, and four marines) traveled from Liverpool via New York to Canada in the spring of 1825 where they were joined by a group of voyageurs. Now Franklin got word that his ailing wife, Eleanor, had died from tuberculosis and that one of his sisters was looking after their baby. His wife of less than two years had urged him to proceed with this expedition despite her ill health. Although grieved by this loss, Franklin proceeded on the trip inland from York Factory.

Franklin's official instructions were to establish winter quarters on Great Bear Lake in a building constructed in advance by the Hudson Bay Company. The spring of 1826, he was to proceed down the Mackenzie River to its mouth, and when the sea ice was open, to map the coast westward to Icy Cape. He would then proceed on to Kotzebue Sound where a ship would be waiting to pick up his party. At the mouth of the Mackenzie, Dr. Richardson, Mr. Kendall, and party would go eastward, mapping the coast to the Coppermine River. Dr. Richardson's task was to gather as much information as possible on the natural history of the region. George Back was to take command of the expedition in case Franklin died.

In June 1825, the entire party began the trip from York Factory to the Great Bear Lake winter quarters. Franklin then took an advance party of his surveyor, Mr. Kendall, an Eskimo interpreter, and a boat crew of six to make a reconnaissance trip down the Mackenzie River. Along the river, Franklin wrote of the now-famous Arctic mosquitoes. "When we landed to sup, the mosquitoes beset us so furiously that we hastily despatched the meal and re-embarked. They continued, however, to pursue us and deprived us of all rest."[2]

Reaching the mouth of the Mackenzie in August 1825, he noted that the sea was free of ice, and seals and orcas were visible in the water. At this point he felt great satisfaction at the progress the expedition was making. Making camp on the beach, Franklin raised a silk British flag, made by his late wife Eleanor, especially to be flown when he reached the polar sea. The group gave three hearty cheers and raised a toast to King George IV. Unfortunately, the brandy reserved especially for the event had been mixed with sea water by one of the voyageurs and could not be drunk. Franklin followed the custom of the early days of exploration and poured the salt-watered brandy on the beach. These acts affirmed Britain's claim on the vast territory. Franklin also deposited letters in a cairn for Captain Parry, who was supposed to make his way to this point after sailing through the Passage. Parry, of course, had been stopped by ice at Melville Island over 600 sea miles to the north. Despite his failure to complete the Passage, Parry achieved fame for sailing a record distance into it.

On September 5, Franklin returned to winter quarters on Great Bear Lake from his reconnaissance to the coast. There were fifty men at winter quarters, which Richardson and Back had named Fort Franklin to honor their commander. The fifty included five officers, nineteen British seamen, marines, voyageurs, nine Canadians, and two Eskimos with family members of three women, six children, and one Indian boy. Their diet depended largely on fish from the Great Bear Lake. Twenty of these people lived at the fishing quarters a few miles from Fort Franklin. In the autumn of 1825, before the lake froze, the fishermen could net between three hundred and eight hundred fish daily. With continued fishing and hunting, the entire party had ample food. Through the winter, the British crewmen had classes on reading and writing taught by the officers. Prayer services were held twice every Sunday.

The officers had assignments to make hourly observations of temperature, magnetism, and atmospheric conditions from 8:00 a.m. until midnight. Each officer also had a special duty. Lieutenant Back worked on finished drawings from the sketches made during their overland travel. Dr. Richardson, the medical officer, took care of the sick and made observations of the natural history. Kendall, the surveyor, drew maps of their travels. During winter they received mail occasionally, thanks to the Hudson Bay Company. Typically the mail was six months old by the time it reached them. For example, on January 16, 1826, a bundle of letters arrived that had been posted in England the previous June, not long after the beginning of the expedition.

By the time they were ready to go to the coast in the spring of 1826, two of the three chronometers had accidentally broken, so Mr. Kendall was given a pocket watch as a substitute to use in the computation of longitude. Fortunately, he also knew how to compute longitude by the lunar distances method, measuring the angular distance between two celestial bodies, such as the moon and a star or the sun, as a means of determining time. Alternating use of the pocket watch with the lunar distances method allowed him to adjust for inaccuracies in the pocket watch. They also used a chip log for measuring distance traveled by boat. Mr. Kendall was skilled in marine surveying and could make an accurate survey of the coast.

They probably used the running survey method for mapping the coast. For such large distances, a careful mapping survey using baselines and triangulation would have been far too time-consuming. The running survey method involves establishing latitude and longitude coordinates at a starting point, then moving the boat parallel to the shore, recording course and distance, and stopping to measuring angles to shore features from different points on the water. The intersection of sight lines to shore features from successive points on the water makes it possible to plot the shore feature with reasonable accuracy. With prominent coastal features established, the coastline between can be sketched by eye. At intervals along the boat's traverse, another longitudinal location is determined to check the positions estimated by dead reckoning.

Franklin noted with satisfaction the vast difference in this expedition from the first one. They were now well equipped with much sturdier boats rather than the bark canoes, and had an ample supply of food. The prospects for success looked very good. Franklin's party had sixteen men in two boats. Richardson's group numbered twelve men in two boats.

When the two parties reached the mouth of the Mackenzie River on July 9, 1826, the sea was still frozen. From their reconnaissance trip the previous August, they expected open water by this date, but they were now delayed until open water appeared. As leads began to open in the water, they made their way, Franklin's group to the west and Richardson's group to the east. Both groups slowly progressed despite constant plagues of mosquitoes. Frequent fog presented another difficulty along this coast, keeping tents and firewood wet.

Franklin's group made its way westward past Prudhoe Bay. As the season was late, Franklin realized they could not make it to Icy Cape, their intended destination. They had now traveled only halfway to Icy Cape from the mouth of the Mackenzie River. His instruction had specified that he should turn back for winter quarters by August 20. At the point where Franklin decided to turn back, which he named Return Reef, a clear sky allowed him to make observations for the first time in over a week. He determined that they had reached 149°37′ W and 70°24′ N. "We quitted Return Reef on the morning of the 18th and began to retrace our way towards the Mackenzie."[3]

They built a driftwood cairn and left a message in a tin box for Captain Parry, who never made it this far. Driftwood had become abundant during a recent storm, making it possible to build fires and prepare hot meals for the first time in days. They also left a message stating how far the expedition had come, hoping the Eskimos might deliver it to the Russians and that the Russians in turn would notify the British government. It is surprising today to realize how much communication depended on the chance that someone would find your message. All this messaging was done in case the expedition never returned to England. They reached Fort Franklin safely on September 21, 1826, to find that Dr. Richardson and his party had already returned from their trip to the Coppermine River. Since departing from Fort Franklin three months earlier, the two groups had traveled 2,048 miles and successfully mapped over 1,000 miles of the Arctic coastline. This was truly a milestone event in mapping the Arctic.

Dr. John Richardson wrote a separate report of his part of the expedition, *Account of the Progress of a Detachment to the Eastward*, which was included in Franklin's *Narrative of a Second Expedition to the Shores of the Polar Sea*. John Franklin received honors for his achievements in the second expedition. He was knighted in April 1829, awarded an honorary doctorate from Oxford University, and received a gold medal from the *Société de Géographie* in Paris. In 1828, he married Jane Griffin. Lady Jane, as she became, was an intelligent woman who was devoted to John and did much to promote his career.

Soon after Franklin returned from his second expedition, he hoped to persuade the Admiralty to let him finish mapping the uncompleted part of the western Arctic coast to Icy Cape, but his proposal was turned down. The expense of launching a major expedition to complete a small segment of the map far outweighed Franklin's desire to finish the job. He had to wait until 1845 until he was given command of another expedition.

14

John Ross's Second Voyage Lasts Four Hard Years, 1829–1833

∽——————————————————————————————

AFTER ELEVEN YEARS on the sidelines, John Ross began making a slow comeback. The Navy's low interest in Ross after his Croker Mountain mistake forced him to seek support from private sources. A wealthy businessman and friend of Ross, Felix Booth, took an interest and agreed to support Ross in finding the north magnetic pole and exploring Prince Regent Inlet for a passage to the west. Although the venture was dangerously underfunded, Ross pushed ahead in 1829 to buy and outfit a ship. He bought the only affordable ship, the *Victory*, a side-wheel steamer of the earliest vintage. It was built for coastal trade and was not designed for major ocean voyages, especially not to the Arctic. The *Victory*'s steam engine was so old that it proved useless. Worse, it took up an enormous amount of space that was needed for supplies and provisions. The designers of this ship expected that the sails would be used for auxiliary power only, so the sail rigging alone was inadequate. The funds were insufficient to buy supplies for the voyage of two years that Ross was planning. Ross turned to the expectation that he could find cached supplies that William Parry had unloaded from his abandoned ship, the *Fury*, in 1825 on his third voyage in Prince Regent Inlet. The site of the food cache became known as Fury Beach. See figure 11.2 in Chapter 11 for the location of Fury Beach and the route that Ross followed.

Few explorers would begin a venture with so many handicaps. What if the supplies cached four years earlier had been washed away or destroyed by storms or polar bears? What if the canned food had spoiled? As it happened, the supplies cached by Parry in 1825 were still in excellent condition and of great quantity. Canned meat and vegetables were as good as the day they were cooked. Bears had visited the cache but had ignored the canned food. The cans held their seals intact so that no odor leaked for bears to smell. Ross later wrote that they all had reason to be thankful to Mr. Donkin. Bryan Donkin, a British industrialist and inventor, had acquired the patent for preserving food in tin cans, and in 1814 he made a contract with the Admiralty to supply the navy with tinned food.

Ross found the tinned food was not frozen and still had good flavor. The cache contained sugar, bread, flour, cocoa, lime juice, pickles, and wine and spirits, all in good condition. The sails from the *Fury* were found dry and ready for use. This was a remarkable find considering the passage of four severe winters with gale force winds and temperatures of -40°F, bears, and Inuit hunting parties. No trace of the abandoned ship was found. Every scrap of wood and iron must have sunk or drifted away. The nearby gunpowder magazine had lost its roof and the canvas cover was in shreds, but the gunpowder was perfectly dry. Ross selected enough gunpowder for their needs and destroyed the rest to prevent possible injury to curious Inuits. The *Victory* now had sufficient provisions for two years and three months, and with additional game from hunts, they could make the food last longer. Their good fortune in finding the large cache of supplies free of damage was almost beyond belief.

As Commander Ross sailed through the ice-clogged Prince Regent Inlet, he wrote an eloquent account of sailing through such hazardous seas, sensing that any moment could bring disaster. Ross was candid about his fear while sailing among dangerous icebergs. In fact, contemporaries criticized him for admitting fear and hopelessness in an era when men did not admit any weakness.

> For readers, it is unfortunate that no description can convey an idea of a scene of this nature: and, as to the pencil, it cannot represent motion, or noise. And to those who have not seen a northern ocean in winter—who have not seen it, I should say, in a winter's storm—the term ice, exciting, but the recollection of what they only know at rest in an inland lake or canal, conveys no ideas of what it is the fate of an arctic navigator to witness and to feel. But let them remember that ice is stone; a floating rock in the stream, a promontory or an island when aground, not less solid than if it were a land of granite. Then let them imagine, if they can, these mountains of crystal hurled through a narrow strait by a rapid tide; meeting, as mountains in motion would meet, with the noise of thunder, breaking from each other's precipices huge fragments, or rending each other asunder, till, losing their former equilibrium, they fall over headlong, lifting the sea around in breakers, and whirling it in eddies; while the flatter fields of ice, forced against these masses, or against the rocks, by the wind and the stream,

rise out of the sea till they fall back on themselves, adding to the indescribable commotion and noise which attend these occurrences.

It is not a little, too, to know and to feel our utter helplessness in these cases. There is not a moment in which it can be conjectured what will happen in the next: there is not one which may not be the last; and yet that very next moment may bring rescue and safety. It is strange, as it is an anxious position; but though fearful, it often giving no time for fear, so unexpected is every event, and so quick the transitions. If the noise, and the motion, and the hurry in every thing around, are distracting, if the attention is troubled to fix on any thing amid such confusion, still must it be alive, that it may seize on the single moment of help or escape which may occur. Yet with all this, and it is the hardest task of all, there is nothing to be acted, no effort to be made: and though the very sight of the movement around inclines the seaman to be himself busy, while we can scarcely repress the instinct that directs us to help ourselves in cases of danger, he must be patient, as if he were unconcerned or careless; waiting as he best can for the fate, be it what it may, which he cannot influence or avoid.[1]

Ross finally decided the steam engine was worse than useless. He realized they were operating primarily as a sailing vessel and gaining little advantage from the antiquated engine. Also, the engine plus its fuel used two-thirds of the available tonnage of the ship. The extra men required to operate the steam engine were experienced with engines but were not trained seamen, thus leaving the nautical crew under manned. His solution to this problem was to dismantle the engine and boiler and store them on shore. The *Victory* was converted to a sailing ship while the ship was frozen in for the winter.

At the beginning of October 1829, the crew began making preparations for winter. They found a secluded bay in the southern waters of Prince Regent Inlet, named Felix Harbor after the benefactor, Felix Booth. The ice was already six inches thick, and the crew proceeded to saw a path for the ship to reach a safe place for the winter. Sailors experienced in the Arctic learned that ships to be frozen in for the winter must locate a spot away from the open water where crushing ice floes and icebergs move about with the wind, tides, and currents. These dangerous movements of ice could crush a ship or lift it out of the water, tilting it so much that it would become uninhabitable. Ross wrote his experience of being frozen to the sea. He described their situation as prisoners with feelings of helplessness and hopelessness, facing many long and weary months of confinement. In an effort to describe the experience of an Arctic winter, Ross wrote that the most imaginative writer would be pressed to describe that which offers no variety, where nothing moves and nothing changes, but everything is cheerless, cold, and still. He described the entire Arctic landscape in winter as a "dull, dreary, heart-sinking, monotonous waste that paralyzes the mind causing one to stop caring or thinking."

Preparing a ship for winter temperatures and winds was a daunting task. They established a powder magazine on a nearby island and moved all the gunpowder to a safe location away from the ship. After removing the engine and some of the supplies, the ship was much lighter but would not rise in the water because it was frozen tight. The crew had to cut away the ice around the ship to allow it to float at its new level, which was nine inches higher. The next step was to create a bank of snow and ice as high as the gunwales around the ship to act as a barrier to the cold and wind. They moved the ship's galley to the center of the men's sleeping berths to more evenly distribute the heat. The deck was covered with thirty inches of snow and tramped down to make a solid mass of ice, then sprinkled with sand to make walking possible. Sails were made into a roof over the deck with canvas extended down to the snow bank at the gunwales. A large copper pipe was extended from outside to bring air directly to the central fireplace. This prevented cold drafts passing through the living quarters. Moisture condensers were devised to prevent a build-up of water in the quarters. The drier air made it easier to maintain a comfortable temperature in the ship with less fuel. Altogether it was a well-planned preparation for winter.

As winter set in and seamen had fewer chores, Ross established a routine to keep the men active. In addition to cleaning and maintenance duties, they were required to walk around the deck beneath the roof and to attend evening school from 6:00 to 9:00. They studied reading, writing, arithmetic, and navigation. All but three of the men could read, but most had poor skills in arithmetic. School work ended with Bible readings and evening prayer.

Ross and the crew made cautious contact with the Inuit in the area. After laying down their guns and shouting the one Inuit word of friendly greeting they had learned from previous voyages, they found the Inuit eager to throw down their knives and advance unarmed. After much embracing and touching, the men gave gifts to the Inuit, invited them on board, and let them taste some of the navy's food, which the Inuit disliked. Ross observed that the Inuit seemed to be better clothed and fed than the crew. A few days later, a number of the ship's crew went to visit the Inuit settlement, even though the temperature had fallen to -37°F. What a different approach toward the local population this was compared to that of Martin Frobisher, who was intent on kidnapping some of the Inuit to take back to England in 1576, resulting in serious battle.

At the end of August 1830, the *Victory* was still held fast in the ice where it had been for eleven months since the previous October. The crew had completely refitted the ship for sailing. However, they found themselves behind a shallow bar, which blocked their passage to open water in a period of unusually low tides. This meant they had to unload much weight from the vessel and try to work it over the bar. Not until September 16 did the *Victory* exit the bay, where it had been imprisoned for nearly a year. Ross wrote of the elation and sense of liberation felt by all of the crew when they were finally freed from the ice and back at sea in control of the ship.

They advanced about 3 miles before meeting a ridge of ice blocking further movement. By September 29, after only two weeks, the *Victory* was again locked in ice and had to be moved into a nearby bay by sawing a path through the ice. This task took nearly a month of hard daily labor, cutting ice that was by now 16 feet thick. By the end of October, they had moved the *Victory* three hundred yards to its new winter position. Again, they had found safety from the crushing ice floes of the open water and prepared the ship for another winter.

The second winter (1830–1831) was much the same as the first, except the men felt demoralized much sooner than before. They still had ample provisions, and no symptoms of scurvy had appeared thanks to the good supply of lime and lemon juice. They again met the local Inuit from time to time and developed a good relationship with the group. The Inuit provided a great benefit to the ship's crew by trading fish for the European's knives. In the spring of 1831, Ross took a party of men with a sledge to explore the west side of Boothia Peninsula. There he discovered the location of the north magnetic pole using a dip meter. He then sledged across a narrow strait to explore and map King William Island. In addition, Ross concluded that Boothia was a peninsula rather than an island. These discoveries, in addition to managing the survival of nineteen of twenty-two men for four years in the Arctic, would assure Ross fame and recognition upon his return to England.

At the end of August 1831, they again managed to free the *Victory* from the ice, and with much effort moved the ship out of the bay. This time they sailed 4 miles, only to find the route again blocked by ice. By the end of September, they were again locked in ice in Victoria Harbor and faced a third winter in the Arctic (1831–1832). This time their plight was more serious. It was time to be concerned about rationing food and preparing to escape overland to an open sea lane. For that purpose, the carpenter began building sledges for carrying their boats and provisions. By the end of November, signs of scurvy began to appear but could still be held in check by the aged lemon juice. One man continued to suffer from the disease in spite of the citrus juice and eventually died.

By January 1832, the crew's medical condition began to deteriorate. Many men reported ailments with no specific disease. When old scars began to bleed, they knew that scurvy was setting in. In April they began preparations to leave the ship and go overland to the site where the *Fury*'s supplies were still stockpiled. The plan was to advance in stages by moving the boats and some provisions partway so they could advance more easily later when they began the trek to the *Fury* beach. This required walking back and forth repeatedly until they moved all the necessary provisions and boats to the first station, 18 miles from the location of the icebound *Victory*. Ross estimated that their walking back and forth three times totaled 108 miles for each man, but their actual advance was only 18 miles. Gale force storms hindered their progress and forced many days of waiting at the ship for weather to clear. When one cache was established, they set out from there to start another farther along and continued this process until all the needed provisions had been moved. This was an elaborate plan that required much walking back and forth between stations. Some men began to protest all the effort and said they wanted to leave behind all they couldn't move and proceed directly to the location of the *Fury*. Ross forcefully ordered them to continue his plan for moving all the provisions.

Finally, by July 1st they reached the beach where the *Fury* had been abandoned. Here they built a shelter 16 feet by 7 feet using lumber cached from the *Fury*'s cargo. They covered the wood frame with canvas and divided the shelter into two rooms to separate officers and men. The space for officers was further divided into four quarters for individual officers. The cooking area was located in a separate tent. All this was completed by the first week of July.

During the remainder of July, the ship's carpenter prepared the *Fury*'s longboats for sails to assist in the rowing. By the first of August, the ice began to break up and navigable water lay ahead. After sailing just 8 miles, ice floes began to threaten them dangerously, forcing them to retreat to the nearest beach, barely avoiding a crushing surge of ice that had formed great ridges near the shore where they had taken refuge.

By the last week of September, after repeated failed attempts to advance over several weeks time, Ross decided the only choice was to return to their canvas-covered shelters on *Fury* beach. This meant yet another winter in the Arctic with food supplies running low. Although the bread ration was gone, the cook could still supply dumplings made with flour and water. They again banked snow around and above their canvas shelter. The temperature in the shelter, away from the side walls, could be maintained at about 45°F. By thickening the snow bank around the shelter and building a floor, they were able to make it comfortable.[2]

By February of 1833, conditions had declined. The carpenter died from scurvy despite the continued availability of lime juice. Ross began to see signs of scurvy in himself as old wounds began to fester. For the first time, it seemed to him that he might not be able to overcome all the difficulties. As spring came, they returned to the difficult task of moving supplies forward a little at a time in preparation for the trek to open water, whenever that might be. By July, all the preserved meat was gone and they resorted to hunting. By the middle of August, they found leads of open water, put the boats in, and finally were underway. Trying to forget they had done all this exact work the year before, they knew they must try again to move ahead.

Their luck continued to hold. On August 17 they encountered open ocean in Prince Regent Inlet, progressing 72 miles into Lancaster Sound, stopping at Cape York on the north end of Baffin Island where they might expect to see a passing ship.

Whalers regularly sailed these waters in summer, so Ross and his crew stayed in the area until a ship appeared. On August 26 at 4:00 a.m., the night lookout sighted a ship. They set a signal by burning wet powder to create a lot of smoke. They launched the longboats and made way toward the sighted vessel. To their dismay, the ship did not stop, but continued on, leaving them behind. At 10:00 they saw a second ship, which slowed due to the calm winds, and Ross's crew were able to come closer and be seen. The ship turned out to be the *Isabella*, which Ross himself had commanded on an earlier voyage. When Ross identified himself and his lost ship *Victory*, the mate of the *Isabella* said that Ross and his crew had been presumed dead for two years.

Ross and his crew were given a hearty welcome. Only then, in contrast to the *Isabella's* healthy and well-dressed crew, did Ross realize how miserable he and his surviving seamen must look. They were unshaven for weeks, wore tattered clothing, and were gaunt and bony. With unimaginable joy and excitement, the crew bathed, shaved, ate, and donned new clothing. In the midst of this activity the *Victory's* crew related the story of their four-year ordeal and escape. In turn, the crew of the *Isabella* updated the others on the news from England. Each of the rescued men expressed his gratitude for deliverance from starvation, scurvy, and death and return to civilization.

After this four-year voyage, John Ross was again in good standing with the Admiralty. He had successfully returned with all but three of the seamen, a remarkable achievement after four years of being locked in the ice. The scientific observation of the north magnetic pole and the exploration of new land by surveying the Boothia peninsula, parts of the Gulf of Boothia, and a great part of King William Island gave John Ross much deserved recognition. In 1834, Ross was knighted, and the geographical societies of London and Paris awarded him gold medals. In 1835, Ross wrote a long narrative detailing the four-year voyage. His account is engaging, thoughtful, and still makes fascinating reading.

GEORGE BACK'S RESCUE EXPEDITION, 1833

When John Ross had failed to return to England by 1832, plans began for a rescue expedition. Because Ross had been instructed to proceed down Prince Regent Inlet, George Back proposed that an overland approach from the south would intercept Ross's route. Commander Back had been with John Franklin on his two overland trips in 1819 and 1825 and had become a well-seasoned land traveler in the Arctic. Anticipating that Ross would have headed for Fury Beach where needed provisions had been cached, Back proposed to head overland through the Canadian Arctic and down the Great Fish River. Back's route is shown in Chapter 11, figure 11.2. From there, the distance to Fury Beach would be no more than 300 miles. The distance from the Great Fish River mouth to the site where the *Victory* was hopelessly locked in ice and abandoned was actually less than 200 miles.[3]

Back was supported by a combination of government and private funds, including the Hudson Bay Company and the Royal Geographical Society. Along with Back, a young surgeon named Richard King, some soldiers, Canadian boatmen, and Indian guides rendezvoused at the Great Slave Lake where they spent the winter at Fort Reliance. There Back reestablished contact with the Indian chief, Akaitcho, who had provided lifesaving provisions for the Franklin expedition in 1821. Now the Indians again hunted to supplement the short provisions in Back's party.

During this time at Fort Enterprise, Back received the message that John Ross and crew had been rescued by ship and had returned safely to England. With the message came a new set of instructions changing the objectives of Back's expedition. He was given a copy of Ross's map with the directive to proceed down the Great Fish River and to conduct a coastal survey on the north coast of King William Island. The Great Fish River proved to be a swift and dangerous torrent with rocks, rapids, and sheer waterfalls. After a month of travel on the river, Back and his party at last reached the sea. The two boatmen, James Sinclair and George McKay, earned Back's high praise for bringing the boats safely down the perilous river to its mouth. There Back faced an important decision, and as usual, ice interfered with plans.

Ice still filled the sea and prevented any boating west to Point Turnagain, and inadequate provisions prevented them from waiting for the ice to melt. The alternatives were to explore eastward, where the water appeared to be more open, or return to England. Although it was still early August, Back chose the latter. His young surgeon, Richard King, urged him to head east in hope of finding if Boothia was a peninsula, as John Ross had concluded, or an island. King later wrote that he strongly objected to Back's decision and had argued with Back on the issue. This public disagreement later cast King as a brash young man and cost him the chance to lead an expedition on his own. Although King's ideas for a follow-up expedition to solve the Boothia problem were practical, the Admiralty regarded him, at age 24, as too inexperienced and unready to command.

One outcome of the expedition was to rename the Great Fish River the Back River. Another result was selecting George Back to lead another expedition. This new effort would sail to Repulse Bay, in the northern reaches of Hudson Bay. From there they would portage across the 50-mile wide neck of the Melville Peninsula to a portion of the polar sea now called Gulf of Boothia. From there they would survey along the coast westward toward Point Turnagain. This instruction assumed that John Ross was wrong and that Back's view of open water east of the mouth of the Great Fish River proved Boothia to be an island—a conclusion charged with wishful thinking. The Admiralty still regarded Ross's view as mistaken because he had been mistaken on the presence of a mountain range blocking the end of Lancaster Sound. Ross's judgment was in question by the Admiralty, and he never regained their full confidence.

Back's ship, the *Terror*, became locked in ice for the winter less than 100 miles from Repulse Bay, and after a near crushing of the hull, two near abandonments of ship, and suffering much damage, the ship barely made it back to Ireland. In two expeditions, Back had surveyed the Great Fish River and located its mouth accurately, but not one mile of new coastal land had been mapped.[4] Nevertheless, he received a knighthood for saving the *Terror* and its crew in the second expedition.

15

Peter Dease and Thomas Simpson Extend the North Coast Map, 1837

A TRADING MONOPOLY was established in the vast unsettled region of Canada in 1670. A group of business associates obtained a charter from King Charles II giving them exclusive trading rights for 1.5 million square miles of land draining into Hudson Bay. The group called themselves the "Honorable Company of Adventurers of England Trading into Hudson's Bay," or the Hudson Bay Company. The Company soon established a network of isolated trading posts as collection points for furs of foxes, bears, otters, and especially beaver trapped by French Canadians and Indians living in the area.[1]

The company agents, called factors, who lived in the trading posts, called factories, became explorers and mappers of the region and, as a consequence, had great knowledge of the terrain over large areas and knew local Indians who could serve as guides. Given this situation, the Hudson Bay Company could have played a major role in surveying the northern coasts of North America. On the contrary, the Company was cautious about letting the outside world and potential competitors have information about the area. Despite the Hudson Bay Company's secretive behavior, another group, the North West Company, defied the charter, invaded the territory, and competed vigorously and often violently. They created their own set of trading posts, some within sight of a Hudson Bay post. This competition interfered greatly with Franklin's first overland expedition, but in 1821, the two companies merged and Franklin had excellent support from Hudson Bay Company for his second overland expedition.

After Hudson Bay Company had secured the territory from outside competition, they became more willing to apply their expertise in travel, exploration, and Arctic survival to mapping the unknown regions of Canada. Their cooperation led to an agreement with the Admiralty to survey and map the north coast by trekking down rivers to the coast and filling in gaps not mapped by Franklin.

Peter Warren Dease (1788–1863), a veteran of many years with the Company, and Thomas Simpson (1808–1840), an eager young Scot, undertook such a mapping mission in the summer of 1836. Simpson was handpicked by his cousin, the Hudson Bay Governor, George Simpson, known as the "Little Emperor." Indeed the domain of the Hudson Bay Company was like an empire, and the Chief Factor, George Simpson, held great power.

Dease commanded the expedition with instructions to continue beyond Return Reef that marked the west end of Franklin's survey and to map as far as Point Barrow. After completing the survey to Point Barrow, they were to move east for the next season and extend the coastal survey eastward beyond Franklin's Point Turnagain. The intent was to connect the surveys of Captain John Ross in the north and Captain George Back at the mouth of the Great Fish River (later Back River). They intended to determine if the Boothia partially mapped by John Ross was an island or a peninsula. Completion of the north coast survey would bring recognition to the Hudson Bay Company and a more positive view of them as a major participant in the Arctic. The routes followed by Dease and Simpson are shown on figure 15.1.

The instructions to the expedition made clear that surveying and mapping were their main objectives and that a written narrative was expected. "The necessary astronomical and surveying instruments are provided, to enable you to make surveys, in which you will be as accurate as possible: and you will be pleased to prepare a full and particular journal, or narrative of the voyage, likewise a chart of the coast: and to take formal possession of the country, on behalf of Great Britain, in your own names, acting for the Honourable Hudson's Bay Company, at every part of the coast you touch: giving names to the different headlands, mountains, rivers, and other remarkable objects you may discover. It is also desirable that you make a collection of minerals, plants, or any specimens of natural history you may fall in with, that appear to be new, curious, or interesting."[2]

Dease proposed that in the first season, eight men would proceed down the Mackenzie River and survey west to the Return Reef, while another group constructed buildings and stored provisions for winter at the east end of Great Bear Lake. In the next season, the party would leave the Great Bear Lake and go down the Coppermine River, then east along the coast to the mouth of the Great Fish River reached by Back two years earlier. They would give names to all features, claiming possession as they went.

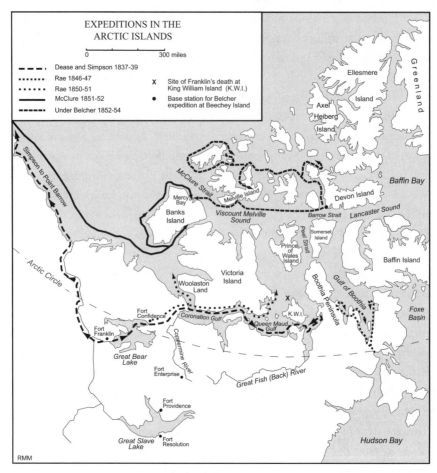

FIGURE 15.1 Many voyages and overland treks were made searching for John Franklin's expedition. A few of the expeditions are represented here to give an idea of the regions they covered.

On June 1, 1837, the expedition left Fort Chipewyan with two boats, the *Castor* and the *Pollux*, after the twins of Greek and Roman mythology for whom the constellation Gemini was named. They headed for the Mackenzie River from Great Slave Lake while the construction party stayed at the lake to begin building quarters, later called Fort Confidence, and make preparations for the next winter. The Company hunters and local Indians began the collection and preservation of caribou and musk ox meat for the food supply.

Starting down the Mackenzie, the mapping party carried thirty bags of pemmican in ninety-pound bags and ten one-hundred-pound containers of flour. By mixing flour with the pemmican to make a thick soup, they could stretch the pemmican. By making this adjustment, the allowance of three pounds of pemmican per man per day could be reduced to two pounds per day. Pemmican is made by drying meat, pounding it to a powder, mixing it with rendered fat, adding a little seasoning if available, and packing it in a can. It could be sliced and fried, crumbled into water as a soup, or, in this case, mixed with flour and water to make a thick, gloppy soup they called "bergoo."

On July 9, the group reached the mouth of the Mackenzie River and soon encountered a greeting party of nineteen Eskimo kayaks. Simpson presented the Eskimos with gifts, but had to fire shots into the air to discourage the Eskimos from following them along the coast as they prepared to set up camp. The explorers traded for fish, which the Eskimos had caught using seines made of interwoven whalebone (baleen).

Strong winds and floating ice impeded their progress, but by July 23, after fifty-three days of difficult travel, the survey party reached Franklin's Return Reef where their own survey would begin. They were ahead of their planned schedule, a feat that Simpson attributed to their persistence in pushing through the softening ice pack, rather than waiting on the beach for it to disperse naturally. A week later, their progress stopped altogether due to intense cold wind and blocking ice pack. At this time, Simpson suggested that he proceed with five men on foot to Point Barrow. Dease agreed to stay with the boats and record tidal observations. Simpson's group packed enough food for the trip and departed with a canvas canoe, a kettle, two axes, some trading trinkets, and the necessary surveying equipment.

Along the way, Simpson encountered a small group of Eskimos who had never seen white men. Through trading, Simpson had managed to acquire one of the Eskimos' large boats, an umiak, and paddles. The umiak held much more cargo than a canoe and could travel in six inches of water. During his time with this group of Eskimos, Simpson found, as had Parry, that they had a very reliable sense of the terrain. One of the women drew a good map of the coast to a point he knew must be Point Barrow, with the number of encampments (days of travel) indicated. The Eskimo map showed a large inlet on the coast. When Simpson's small group arrived at the inlet, Simpson named it Dease Inlet, and they rowed across its mouth in a freezing fog so dense they had to rely on compass navigation alone. He reported his great satisfaction with the performance and navigability of the Eskimo boat.

On August 4, Simpson saw a projection of land pushing toward the sea, and he knew that he had reached his destination, Point Barrow. After hours of pushing through heavy ice, he reached the point, planted the flag, and with cheers from the men, claimed the land in the name of his Majesty Edward IV. Although unknown to Simpson in this remote location, Edward IV had died about six weeks earlier, and Victoria was now monarch. Simpson and company were then confronted by about one hundred Eskimos who seemed not to know they had just become British subjects. The Eskimos appeared to have hostile intent, and the group of surveyors took care to guard themselves and their possessions. The Eskimos' unfriendly demeanor induced Simpson's party to leave the next day, but not before taking the necessary astronomical bearings to get an accurate location for Point Barrow—their very reason for being there.

Strong westerly winds sped them eastward on the return to the Mackenzie River. On passing the Colville River, Simpson noted a strong flow from the river making their progress difficult. Even 12 miles offshore he found the water to be fresh. The survey party met other groups of Eskimos, some of whom had iron cooking pots and knives showing that the trade pattern among the Arctic people extended into Russia and beyond. One of the Eskimo groups was particularly unfriendly, causing Simpson to note how glad he was the Eskimos lacked fire arms.

On August 17, the entire party reached the mouth of the Mackenzie River after little more than five weeks' travel to Point Barrow and back. They returned up the Mackenzie to the newly constructed winter quarters at Fort Confidence. By mapping the coast from Franklin's Return Reef to Point Barrow, they had accomplished the first stage of their mission and filled an important gap in the map of North America and a segment of the Northwest Passage. Thomas Simpson apparently felt quite full of himself over this achievement when he wrote his brother, also a Company employee, that he alone had the well-earned honor of tying the Arctic to the Pacific and posting the British flag on Point Barrow.

In the winter of 1837–1838 at Fort Confidence, Simpson spent his time writing and charting the results of his journey to Point Barrow, hunting, fishing, and reading from the library brought from Hudson Bay the previous spring. He again wrote to his brother describing Dease as a worthy soul, though rather indolent and illiterate. This totally erroneous and undeserved comment shows Simpson's growing ego and self-confidence with the wilderness and his impatience with Dease's leadership of the expedition. Despite this, Simpson and Dease maintained a cordial relationship throughout.

Thomas Simpson's experience in the Arctic was limited and had not yet included life-threatening situations outside the protection of the winter quarters, such as unexpected winter storms and starvation. After his recent success for which he took full credit, Simpson felt he could to anything and assumed that Peter Dease, after many years in the Arctic, was too cautious or lazy. Perhaps Simpson gave his ego free rein because his cousin George was the Governor of the Hudson Bay Company North American enterprise and a strong and forceful presence in the Company.[3]

Despite Simpson's comments, Dease remained in command of the expedition. Dease was, after all, one of the Chief Factors in the Company, which was not a position given to lazy, illiterate employees. His long experience in organizing and directing expedition parties provided the base of support for a successful expedition, which

Simpson had failed to mention. To Simpson's credit, however, his frustration was expressed only to his brother Alexander, and he maintained respectful and amiable relations with Dease.

Thomas Simpson took credit for planning the expedition of the coming spring and assured his brother that his experience and foresight would bring success. In late March, with temperatures still -60°F, Simpson and a small party with dog sledges began a series of three trips transporting provisions to a cache about 95 miles along the trail and only 15 miles from the headwaters of the Coppermine River. In this experience, Simpson encountered some of the effects of Arctic winter and commented that he had to abandon his woolen clothing in favor of more suitable clothes and bedding made from moose and caribou hides.

THE EASTWARD SURVEY, 1838

On June 6, 1838, Dease and Simpson departed Fort Confidence overland and headed back to the Arctic coast via the Coppermine River. At this early stage of snow melt, the Coppermine was still in bankfull stage with many powerful and dangerous rapids. Some rapids were flanked by sheer cliffs, making portage or towing from shore impossible. Parts of the river were still icy or had floating ice blocks, adding to the hazards of the journey. On July 17, the party reached the mouth of the Coppermine and began the eastward coastal trip along the south coast of Coronation Gulf. Floating ice and land-locked ice impeded their travel, and only by hard work could the party push their way through. Ice could squeeze the boats between two floes like scissors, gouging the sides.

Their frustration with ice in Coronation Gulf is typical of the vagaries of ice in the Arctic. When Franklin traveled this coast in 1821, he reported open water and easy travel. In 1838, Dease and Simpson struggled to make progress through the ice-clogged water. Goals for the expedition were set in the expectation, or hope, that ice would not block the way. Now by August 20, after a month of travel along the coast, they had not yet reached Franklin's Point Turnagain, about 200 coastal miles from the mouth of the Coppermine River. By late August, the first signs of winter appeared and the expedition had to consider returning to Fort Confidence. Simpson again took a party of seven men and continued on foot in an effort to achieve at least a portion of their objective, and Dease remained behind with the boats. Simpson agreed to return to Dease's location in eleven days, by August 31.

Simpson reached Point Turnagain and could see a large body of land, which he named Victoria Land (now Victoria Island), about 20 miles across the narrow waters. As he progressed to the east, he feared he might see the water body end in a bay, but to his great pleasure he found that the sea continued. From there, the coast trended to the southeast from his vantage on Cape Alexander, which he named for his brother. This happy view stimulated him to push on eastward along the coast until September 25, when he formally took possession in the name of her Majesty, Queen Victoria. They built a cairn wherein he placed an account of their journey that had successfully surveyed an additional 100 miles beyond Franklin's Point Turnagain. He met Dease on his return, cached the boats for future use, and together on October 14 they returned to Fort Confidence.

The following spring, Simpson received a commendation from the Admiralty for his survey to Point Barrow, and the Royal Geographical Society awarded him a gold medal. This recognition spurred him to propose a third trip to finish the survey that had been cut short the previous summer. The route was already known, boats were cached at the Coppermine, and the Chief Factors of the Hudson Bay Company gave their approval.

THE THIRD SURVEY OF SIMPSON AND DEASE, 1839

In 1839, Dease and Simpson faced the good fortune of milder spring and summer temperatures. They descended the Coppermine River with fewer ice hazards and arrived at the mouth earlier than the previous year. On reaching Coronation Gulf in mid-July, they were pleased to find much open water, permitting easy travel by boat. The expedition again passed the Turnagain Point and continued to the point where Simpson had stopped mapping ten months earlier. The expedition continued eastward by boat, mapping and naming features of the coast as they went. Traversing along the coast of a gulf, later named Queen Maud Gulf, Dease and Simpson reached the Adelaide Peninsula, which they had believed connected to King William Land (now Island). To their surprise, a strait, first named Dease and Simpson Strait (now Simpson Strait), separated the mainland from King William Land, showing King William Land to be an island. Furthermore, the strait connected to an open sea that included the mouth of Great Fish River (now Back River), which George Back had reached in 1834.

Two boatmen with Dease and Simpson, Sinclair and McKay, had accompanied Back to this point in 1834. When they recognized the location, they uncovered a cache left there five years earlier, with gunpowder, fish hooks, chocolate, and two bags of pemmican, all in good condition. Since both Back and Franklin had visited this point on separate expeditions, Simpson was able to compare his longitude and latitude observations. He found agreement with Franklin and a difference of 25 miles with Back.

Because the sea continued to the north, Simpson felt sure it led to the east shores of Ross's Boothia. The question Simpson now wanted to answer was whether Boothia was a peninsula or an island. This was an important question for navigation in the area. If Boothia were an island, passage south of it would lead certainly to Prince Regent Inlet and Lancaster Sound, or to Hudson Bay through the Fury and Hecla Strait. Dease and Simpson had met their objective by reaching the mouth of the Great Fish River and tying together more than 1,600 miles of North American coast line. There remained a tantalizing bit of unmapped land between them and the farthest point reached by Ross, less than 60 miles to the north, and the elusive answer to the question of the insularity of Boothia.

In late August, it was time to begin the long return journey to Fort Confidence. They made a short detour en route across the strait to Victoria Island where they mapped and named details of the coast. As they returned to the mainland in mid-September, they discovered an abundance of driftwood on the beach. This caused the entire party great excitement, as they had not found enough wood in this treeless environment for the luxury and comfort of a big fire since July. On September 16, they again reached the mouth of the Coppermine River. There Dease found a pair of Eskimo-made boots hung on a pole awaiting him. The previous July, Dease had arranged with some Eskimos to make a pair of boots for him, and he was pleased that they had followed through on the bargain. On September 24, they reached Fort Confidence, and all members of the party went their various ways.

Immediately, Simpson wrote a proposal to complete the unfinished segment of the journey between their recently finished survey and the Fury and Hecla Strait. By doing this, Simpson planned also to solve the question of Boothia. Simpson waited through the winter months and received no word. When the spring thaw came, he gave up waiting and decided to return to England. Not long after Simpson left Fort Confidence, the news arrived that his proposal had been accepted. Unfortunately, Simpson never found out because he died on the way home at age 32. The cause of his death is unclear. As he traveled through the United States on his way to England, he had an altercation with men believed to have been with him in the Arctic.[4] Authorities never resolved his death, claimed variously to be murder or suicide.

16

John Franklin's Last Expedition Becomes the Failure

of the Century, 1845

AFTER HIS 1826 expedition, John Franklin received no immediate assignment and along with many unassigned seamen, went on half pay. Naval officers dreaded this kind of standby duty—it essentially put their careers on hold. In 1836, Franklin received the stopgap position of Lieutenant-Governor of Van Damien's Land (now Tasmania) and remained at this post for seven years.

In 1844, Sir John Barrow, Second Secretary of the Admiralty, promoted another search for the sea passage of the North American Arctic. Barrow's argument was that great progress had been made since the push began in 1818. At that time, the only known points on the North American Arctic coast were at the Mackenzie River mouth, the Coppermine River mouth, and Icy Cape in Alaska. By the time of Barrow's proposal, other naval and Hudson Bay Company men had mapped much of the north coast, and many of the larger islands had been partially mapped. A large area of unexplored coast still remained, Barrow posited, between the locations reached from the east by Parry, Ross, and others, and coasts in the west and south reached by Franklin and his parties. Barrow, nearing his 80th birthday and about to retire, felt some urgency to fulfill his long-held goal to find the Northwest Passage. He proposed that the connection could be made by sailing south and west from the area of Cape Walker. They would cut through the unexplored area, and the Northwest Passage would be complete. Lord Haddington, First Lord of the Admiralty, accepted the proposal forthwith.[1]

Barrow was highly respected and had originated many of the past expeditions, so rejection of this proposal was nearly impossible. How could Franklin not accept

Barrow's proposal? Barrow made it sound so obvious, perhaps even easy: just sail from the meridian of Melville Island to Bering Strait, a distance of 900 nautical miles. His argument seemed to ignore the possibility of land or ice lying between these two points. But of course, a proposal must sound convincing. Barrow also argued that English sailors, from Elizabethan times until 1844, had made almost the entire map of the American Arctic. If some other country (meaning Russia, which was also keenly interested in expansion into America) should now finish the job, England would be ridiculed for her hesitation and neglect of the task. This appeal to English pride was no doubt the clincher in selling the proposal. Barrow concluded by writing that the project could probably be finished in a year at little risk to ships or men, saying that in spite of the dangers and hardships of past Arctic voyages, there had been little loss of life. Barrow further persuaded that Arctic voyages had proven to be very good training for sailors of the Royal Navy. This aspect was, in fact, one of the main reasons for renewing exploration after the naval battles of the Napoleonic Wars and very persuasive. Although not in error, Barrow's proposal was a bit oversimplified in the main objectives. Nevertheless, any naval officer would be eager to command an expedition that might be the final breakthrough to the elusive Northwest Passage. Even if the expedition failed to find the Passage, the major advances achieved through the scientific observations—especially additional measurements of the magnetic field—would make the effort worthwhile. Global measurements of the earth's magnetic field, specifically the variation between magnetic north and true north, were a valuable aid to navigation, and additional data was important. European mariners of the sixteenth century knew little of magnetic variation until they began to sail westward.

Because of Franklin's experience in the American Arctic, the Admiralty chose him as one of the officers to help evaluate the proposal. Franklin strongly suggested that the ships proposed for the expedition, *Erebus* and *Terror*, be refitted with steam engines as an aid in pushing through ice-choked waters when winds were calm or adverse. He claimed, with questionable optimism, that this could be done without destroying the ships' capacity for stores and provisions.

Ultimately the proposal reached the Prime Minister, Sir Robert Peel, for final approval. After much discussion, the Admiralty decided that Franklin, age 58, should lead the expedition and command the *Erebus*, and that Captain Francis Crozier, age 48, should be second-in-command of the expedition and in command of the *Terror*. Crozier had experience sailing in both the Arctic and Antarctic and was a highly respected naval officer. Several other officers had more experience sailing in the Arctic than Franklin, but Franklin was well known for his ability to carry out an expedition that emphasized the well-being of the men. Also, Franklin was known to have his men's respect and affection. Other naval officers with Arctic experience were considered. The Admiralty had approached James Ross to command the expedition, but Ross squelched the suggestion, saying that at age 44 he was too old, and that he had promised his wife to take no more expeditions. Francis Crozier also declined to be considered for command of the expedition. With William Parry in retirement, all attention focused on Franklin as the leading choice to take command, despite serious questions about his age and physical fitness for such a rigorous assignment.

The two ships were outfitted with steam for auxiliary power with twenty horse-power railroad engines and screw propellers that could be raised out of the water when the ships were under sail. Because there was little room for storing both coal and provisions, they carried only twenty-five tons of coal, enough for about twelve days of use. This limitation of fuel assumed that the engines would be used only for short periods when needed to navigate in ice. The engines were mounted crossways in the hold behind the mainmast and took up the entire width of the ship. On initial trials in the Thames River, the boats traveled 4 miles per hour. To help compensate for the space lost by engines and coal storage, a transport ship, *Barretto Junior*, would go with them as far as possible.

HMS *Terror* (launched 1813) and HMS *Erebus* (launched 1826) were sister ships built as bomb vessels with a heavy construction designed to withstand the force of firing heavy mortar shells onto shore fortifications. The *Terror* took part in the bombings of Fort McHenry in the Battle of Baltimore in 1814. These two ships were converted to exploration ships and used in several Arctic and Antarctic expeditions before adding steam engines and iron plate reinforcement on the hulls in 1844 in preparation for the Franklin expedition.

Despite Barrow's optimism that the expedition might take only a year, the ships were provisioned with enough supplies to last for three years. They provided for each man to have a daily allowance of one pound of biscuit, two and a half ounces of sugar, one quarter ounce of tea, one ounce of chocolate, and one ounce of lemon juice. Two days a week each man would get salt beef, two days salt pork, and three days a half pound of canned meat. In addition, each ship carried a library of about 1,200 books, including nautical journals, novels, and humor magazines, such as the popular *Punch*. The supply of canned vegetables, soap, candles, stationery, and many other items created a logistics nightmare of organizing and loading. In all, there were 129 officers and men aboard the two ships, with provisions and supplies for three years. Lemon juice, canned vegetables, and canned meat rather than pickled made a significant difference in the diet at sea compared with Frobisher's voyages in the late sixteenth century. The inevitable sea biscuit steadfastly remained, usually infested with weevil larvae and baked months ahead by naval bakeries. French ships reportedly carried flour and baked their bread on board.

The *Erebus*, *Terror*, and *Barretto Junior* sailed down the Thames on May 19, 1845, to sail on a course to be determined by the position and extent of the ice toward the Bering Strait. If passage westward was blocked, Franklin had authority to sail north up the Wellington Channel. It was the most elaborately planned and provisioned expedition ever to search for the Northwest Passage. Though they did not sail through the Passage, it did become the last effort of the British Admiralty to find the elusive northern route to the Pacific. The exploration of the Arctic had captured the interest of the British public, but this expedition, perhaps more than any other, carried the aspirations and best wishes of all Britain. The men of the expedition had complete confidence in Franklin because of his past experience and his good rapport with his officers. The ships had everything deemed necessary for a successful voyage, and failure seemed impossible.

In Disco Bay of eastern Greenland during the first week of July, crews transferred supplies from the *Barretto Junior* and sent the transport ship home, along with reports to the Admiralty and personal mail. Near the end of July, two British whaling ships reported sighting the *Erebus* and *Terror* progressing westward and in excellent condition. The ships and men of the Franklin expedition never returned.[2] Their route is shown in Chapter 11, figure 11.2. The site where remnants of Franklin's expedition were found are shown in figure 15.1, which also lays out some of the searching expedition routes.

Only after years of repeated searches was their fate determined. Lady Jane Franklin became a driving and persistent force in continuing the search for Franklin, and she worked diligently to promote his reputation as the man who had completed the Northwest Passage.[3] However, even the unsuccessful search expeditions bore fruit. Their search missions produced maps of most of the remaining unknown islands of the Canadian Arctic archipelago. During more than ten years of searching for Franklin, the state of mapping in the Arctic advanced more than at any comparable time.

Interest in the whereabouts of Franklin's ships continues to present times. In the summer of 2011, the remains of *Erebus* and *Terror* were objects of an ongoing three-year search by personnel of Parks Canada near the shores of Prince William Island. A team of scientists equipped with side-scanning sonar, multi-beam bathymetry, and an autonomous underwater vehicle searched Victoria Strait west of King William Island.

PART V

The Franklin Searchers Almost Finish the Map, 1847–1858

Had we lived, I should have had a tale to tell of the hardihood, endurance, and courage of my companions which would have stirred the heart of every Englishman. These rough notes and our dear bodies must tell the tale.

—ROBERT FALCON SCOTT

They change their clime, not their frame of mind, who rush across the sea.

—HORACE

17

The First Searchers Look in the Wrong Places, 1847

JOHN ROSS EXPRESSED his misgivings about the suitability of the *Erebus* and *Terror* before Franklin departed in 1845. Ross's experience with Arctic survival led him to believe the two ships were too large, the drafts too deep (19 feet), the crews were too large (a potential problem if short rations became necessary), the steam engines and coal added too much weight, making them ride low in the water, and the steam engines took up space that could better have been used for provisions. Ross's convictions on these issues prompted him to advise Franklin to leave frequent cairns along the way with notes describing his intended direction of travel, and to leave food caches at intervals in case he should lose the ships and have to walk out, as Ross had done. Further, he told Franklin that he would be prepared to lead a rescue party if Franklin's whereabouts were not known by February 1847.[1]

No one, including Franklin, took John Ross's concerns seriously. The expedition had the most detailed planning and preparation of any before that time. Optimism ruled everyone's thinking. When February 1847 came and no word had been heard of Franklin, Ross quickly offered to form a search and rescue expedition. The Admiralty flatly rejected the offer after only a few minutes of discussion. The Admiralty felt the fail-proof expedition needed more time, but no doubt their dislike of John Ross influenced the hasty decision. Dr. William King, who had made overland treks with Franklin, proposed an overland search down the Great Fish River and along the west coast of Boothia, but the Admiralty again rejected the idea. In retrospect, this would have taken him very near Franklin's actual location. By the end of 1847, the

Admiralty began to feel that all was not well. In December they accepted an offer to lead a search mission, not from John Ross, but from his nephew, James Ross, who was highly regarded in the Admiralty.

In the spring of 1848, a large search and rescue project began from three directions. Two ships, the *Enterprise* under James Ross, and the *Investigator* under Edward Bird, approached from the east into Lancaster Sound. Another ship, the *Herald*, under the command of Captain Henry Kellett, was sent around Cape Horn to approach through Bering Strait. Richardson and Rae traveled overland down the Mackenzie River and eastward along the coast looking for signs of Franklin. They found nothing.

In November of 1848, Sir John Barrow, the influential Second Secretary of the Admiralty, died at the age of 84. The Lords of the Admiralty, paralyzed without Barrow to advise them on questions concerning the Arctic, nearly dropped the whole issue of finding Franklin. Royal Navy officers, with professional pride in looking out for their own, brought pressure on the Admiralty to begin a search. Also, concern from the general public was running high. These sentiments pushed the Lords into action. They began by offering prizes of £20,000 for anyone who rescued Franklin, £10,000 for finding his ships, and for good measure, £10,000 for the first person to sail the Northwest Passage. That effort exhausted the venerable Lords' abilities, and they gave authority to the Arctic Council for planning and execution of the rescue.

The Arctic Council, first begun by Barrow, had many qualified men with experience in the Arctic, including, among others, William Parry, James Ross, Frederick Beechey, and John Richardson. Since John Franklin's disappearance, his influential wife, Lady Jane Franklin, also sat on the Council. After Barrow's death, his son John Barrow Jr. became chairman and coordinator for the Council. The Lords of the Admiralty came to rely fully on the advice of this Arctic Council.

By the spring of 1850, a veritable armada of ships began to look for Franklin. Two ships, the *Enterprise* under Captain Richard Collinson and the *Investigator* commanded by Captain Robert McClure, approached the Arctic from the west through Bering Strait. Thirteen ships made an eastern approach into Lancaster Sound, although not all were sent by the Admiralty. Lady Jane Franklin, along with private subscriptions, financed the *Prince Albert*, and the aging veteran John Ross commanded two small ships, for which he personally raised money. Two American ships joined the fleet—one, the *Advance*, carrying well-known explorer/physician Elisha Kane.

They all had only one bit of information to direct the search—the instructions Franklin received before leaving England. He had been directed to sail west toward Melville Island, thence southwest. Should those routes be impassable, he was to go north from Lancaster Sound up the Wellington Channel. This last instruction suggests that the hope remained alive for an ice-free polar sea. The number of ships involved allowed for the exploration of several possibilities at the same time. They attempted to sail south in Peel Sound but found that passage blocked by ice. Actually, when Franklin passed through Peel sound in 1846, it was open. Such is the difficulty of sailing in the Arctic, and searching for a lost ship years after its disappearance held little possibility of success. Refer to figure 15.1 in Chapter 15 for some of the routes followed by the searchers.

On August 23, 1848, one ship, the *Assistance*, commanded by Captain Ommanney, found a rock cairn on a high promontory along the shore of Beechey Island. They also found remains of fires, tracks from sledges, empty food cans, and trash heaps presumed to be the site of Franklin's stay at Beechey Island during the winter of 1845–1846. A few days later, several ships converged on Beechey Island, where the searchers found graves inscribed with names and dates of three sailors known to have been on Franklin's expedition.

Although the searchers learned that Franklin had indeed come this way and had stopped at Beechey Island for the winter, there were still many unanswered questions. They found no note or arrow indicating the direction Franklin had taken after leaving Beechey Island. The purpose of rock cairns was to leave written information about the current situation and intended travel direction. The cairn on Beechey Island was disassembled in a thorough search for a message, but nothing was found. The enigma of the empty cairn defied explanation. They found some sledge tracks heading north about 40 miles along the east coast of the Wellington Channel. This suggested that Franklin was making a reconnaissance of that channel as a possible route for the next summer. Perhaps this increased some searchers' inclination to look for Franklin in that direction. The searchers surmised that Franklin perhaps had tried to sail farther west or south but, finding the way blocked by ice, had sailed north through Wellington Channel.

Another question was the cause of death of Franklin's three crewmen. John Ross had been lost for four years with only one crewman's death. Franklin's first overland expedition was totally without rations, and although ten of the French Canadian voyageurs died, none of the naval personnel died from starvation or scurvy. Franklin, "the man who ate his boots," had received much recognition for bringing his men home alive. However, the 1845 voyage had been prepared for three years at sea. What could have gone wrong in the first year? These questions must have crossed the searchers minds, but they found no clues to suggest an answer. When the sailing season ended, the search armada stayed for the winter, except for Lady Franklin's ship, the *Prince Albert*, which returned to England with reports of the discoveries at Beechey Island.

An important change now took place in the method of search and mapping. Rather than stopping when ice blocked the way, now manhauled sledge teams spread out across the Arctic unimpeded by frozen seas. This was not the first time for using sledges, but it had now become the primary means of transport once the ships met their limits and froze into a winter harbor. A team of six to eight men harnessed to a heavy sledge loaded with eight hundred pounds of cargo could average 10 or 12 miles per day under favorable conditions. If they made a major effort, they could travel as far as 20 miles in a day. Usually the officers did not take the harness, but carried instruments and a telescope. The sledges carried food supplies, tents, and often a boat for crossing stretches of open water. The strenuous exercise of pulling sledges consumed a great number of calories and required a great increase of rations to make it successful.

During the spring of 1851, Lieutenant Leopold M'Clintock led a group of sledges from the winter station near Beechey Island some 380 miles to Melville Island where William Parry had harbored for the winter of 1819–1820. This was one likely route for Franklin to have taken, but M'Clintock found no trace of the expedition, nor did he map any new territory. On the return journey, M'Clintock's group of sledges considered the prospect of going down Peel Sound, but decided against that direction. Again Franklin's actual route was bypassed as unlikely.

In the summer of 1851, the fleet of ships returned home with a completely untrue rumor, begun by John Ross's Eskimo interpreter, that Franklin's expedition had all been killed in an attack by Eskimos.[2] This rumor was believed by some in the search parties, but John Ross himself gave it no credit. Nevertheless, he did not effectively squelch the rumor and was vilified by Lady Franklin for relating it. Lady Franklin was upset that the rumor cooled the Admiralty's interest in continued searches. Poor John Ross got no respect from anyone, even when he did well and at his own expense. He retired from naval duty at age 73 with a debt of over £500 incurred by the fruitless rescue expedition.

The Admiralty's interest flagged, but the public cried out for an answer to their hero's whereabouts. Interest in the Northwest Passage was not the issue. That quest suddenly seemed unimportant compared to the losses of men and money incurred; the quest for the lost hero became foremost. Faced with this strong sentiment, the Admiralty again directed the Arctic Council to devise another search operation. What could the Council now do? Their past expeditions had given little attention to the probable route that Franklin might have taken. His orders directed that Wellington Channel be tried as an alternative only after other routes to the south or west had failed. The expeditions of 1850–1851 had given considerable attention to Wellington Channel, with few probes to the south, which had been Franklin's first designated direction to explore. Seeing Peel Sound choked with ice probably caused the searchers to assume that it had also been choked with ice when Franklin saw it in 1846. One member of the Arctic Council earlier suggested sending a sledge party all the way down Peel Sound to the mouth of the Great Fish River. Had that been done, Franklin's ships almost certainly would have been found. Another member of the Council, for no particular reason, dismissed the idea that Franklin would have gone that far south. Lingering belief in the open polar sea swayed the group to direct searches northward up Wellington Channel and other north trending routes. The various misguided paths the searchers followed resulted in an immense addition to the known coastlines. Figure 17.1 shows the amount of known areas added as a result of the search for Franklin.

The outcome of this debate was another fleet of five ships being sent on a search mission in the summer of 1852 under the command of Captain Edward Belcher (Chapter 15, figure 15.1). The ships were the *Assistance*, commanded by Belcher, the

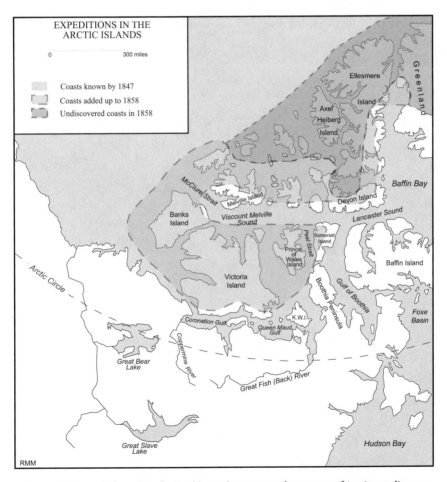

FIGURE 17.1 At the end of searching for Franklin explorers mapped a vast area of Arctic coastlines as shown by the intermediate shading. The darker shading shows the remaining unmapped area at that time.
Sources: Berton, *The Arctic Grail*; and Neatby, *Search for Franklin*.

Resolute under Captain Kellett, the *Intrepid* with M'Clintock, and the *Pioneer* under Sherard Osborn. The *North Sea* went as a supply ship for the fleet. Some bizarre and totally futile efforts were made to contact Franklin. They trapped Arctic foxes and released them with collars containing information intended for Franklin. Medals with similar information were given to Eskimos in the hope that Franklin's party would see them. Balloons with messages drifted over the landscape. Cliffs facing the sea were painted with information visible from a distance. Sledge parties, however, remained the primary effort. They focused on Wellington Channel and mapped many uncharted coasts, but they were well north of Franklin's planned route. Captain Belcher received orders to send two ships west to Melville Island to search for lost ships commanded by Captain McClure and Captain Collinson who had sailed to the Arctic in 1850 via the Straits of Magellan and Bering Strait and had not yet returned. Now a search for the lost searchers had been added to the assignment. The contributions of the McClure and Collinson expeditions merit a closer look.

18

John Rae Hears about Franklin from Eskimos, 1848

BY THE END of 1846, people closest to John Franklin began to feel concern about the fate of his expedition. Many felt he should have returned or sent word of his location after two years. In early 1847, Franklin's friend and companion on overland expeditions, Sir John Richardson, wrote to the Admiralty. He proposed caching a large amount of pemmican at intervals along the Arctic coasts of North America for Franklin and his company. His presumption was that Franklin would try to go overland to a Hudson Bay Company outpost if his expedition had become hopelessly stopped. The Admiralty agreed to this and ordered a naval stores facility to manufacture more than 17,000 pounds of pemmican. They also ordered four boats suitable both for cargo and overland portages.[1]

In September, twenty men with equipment and provisions arrived at York Factory on the west side of Hudson Bay. Arriving so late in the season forced the party to stop for the winter of 1847–1848 at the Hudson Bay Company outpost, Cumberland House, about 430 miles inland. Richardson remained in England making further arrangements. See Rae's route of travel in Chapter 15, figure 15.1.

When no word of Franklin had come by the end of 1847, concern began to increase within the Admiralty. At Richardson's suggestion, Dr. John Rae, a Chief Trader of the Hudson Bay Company, was appointed as second officer. Rae's Arctic experience and Hudson Bay Company connection made him a valuable addition to the expedition. Rae had proven himself a capable explorer and was adept at living off the land. In 1847, Rae had made an important contribution to the Arctic map by mapping the south end of the Gulf of Boothia, called Committee Bay, and demonstrating that John Ross had been correct in his assertion that Boothia was a peninsula, not an island. By this accomplishment, done entirely on foot, Rae mapped 625 miles of new coastline. Despite Rae's determination that Boothia was not an island, John Barrow of the Admiralty persisted in the assumption that there would be a passage between Boothia and the mainland. Barrow could not accept that John Ross might be right about something.

Although the search for Franklin was first priority, the Richardson-Rae expedition would add much information about the Canadian Arctic. Richardson, a surgeon and published naturalist, carried meteorological and magnetic instruments as well as supplies for collecting and recording plants. Richardson and Rae arrived in New York in the spring of 1848 with about four thousand pounds of baggage and supplies, including maps and the usual mapping tools—sextants and chronometers. Although Richardson selected Rae as second-in-command, the two men were not particularly well matched. Richardson was then 60 years old and could not equal the much younger Rae's endurance. Rae wrote of annoyances between them relating to the slower pace required for Richardson. Rae also felt concern for the lack of Arctic experience among most of the crewmen. They were not used to carrying heavy loads, and Rae frequently caught them trying to ditch part of their loads. He felt that Hudson Bay men could have served the expedition better.

Richardson and Rae were trekking overland in the Arctic in 1848 concurrently with two other search expeditions at sea. Commander T. E. L. Moore and Captain Henry Kellett, sailing the *Plover* and *Herald*, made a western approach through Bering Strait to search the Alaskan shore eastward to the Mackenzie River. The other was Sir James Ross in the *Enterprise*, accompanied by Captain Edward Bird in the *Investigator*, sailing into Arctic waters from the east. Each of these expeditions carried extra provisions for Franklin's expedition, should they find them. The Admiralty, while responding to the concern others had for Franklin, still felt it was probably too soon to think that Franklin's well-planned and well-provisioned expedition had faltered.[2]

Richardson had planned to reach the mouth of the Mackenzie by the end of August and begin his coastal search. Although it was unknown what lands lay to the north, Richardson expected to explore other areas opposite the coast of the mainland known only as Woolaston Land and Victoria Land (Victoria Island). It was now autumn and too late to continue, so they spent the winter at Fort Confidence on Great Bear Lake. The following spring, Richardson dropped out of the expedition and returned to England. Rae's efforts to reach Victoria Land were blocked by ice, and he went to Fort Simpson on the Mackenzie River for the next winter. In the end, the expensive, slow-paced expedition found no trace of Franklin and had mapped no new coastline. However, John Richardson, with his fellow naturalists on the expedition, made a great contribution to the natural history by publishing two volumes on the flora and fauna of the Canadian Arctic.

Five years later, in 1853, search parties were still looking for Franklin. Rae returned to the Arctic to examine the south end of the Gulf of Boothia. He approached from Repulse Bay at the north end of Hudson Bay and the following spring crossed an isthmus of land now named for him. He continued across Boothia Peninsula, showing it to be a peninsula, and crossed to King William Island, proving it was not connected to Boothia. When he reached Pelly Bay in April, Rae met Eskimos who told him of thirty to forty white men who had starved to death several years earlier at a location farther west. Rae remarked that his knowledgeable informant wore a band around his head that came from the place of the dead men. Rae bought the Eskimo's headband and told him he would also buy any other relics from the site of the dead men. Rae could not determine where the items came from or if the starved men had been part of the Franklin expedition. However, he did not doubt that he had found the first clue to the Franklin party's fate. Who else could it have been? Rae believed in the Eskimos' truthfulness, having no reason to think this Eskimo would make up such a tale.[3]

Rae later said he did not pursue the lead about Franklin because the terrain was still covered with snow. Also, his first interest on this expedition was to continue mapping the Arctic coastline. He had made important discoveries distinguishing between islands and peninsulas in the Boothia region and was intent on finishing that job. He was, after all, a Hudson Bay Company employee and had been instructed to map rather than find Franklin. When Rae reached the west side of Boothia Peninsula, he encountered a body of water now called Rae Strait. The existence of this strait proved that King William Island was separate from Boothia Peninsula. Also, the presence of young ice in that strait indicated that the water was ice-free for a few months of the year, and that the strait was protected from the crush of pack ice that flowed into Victoria Strait on the west side of King William Island. This was an important discovery. "John Rae knew he had discovered, running between Boothia Peninsula and King William Island, the hidden gateway link in the Northwest Passage. This was it!"[4] The Norwegian explorer, Roald Amundsen, proved Rae was right by traversing through the entire labyrinth of the Northwest Passage in 1903–1905.

Later, when Rae returned to Repulse Bay in the autumn of 1854, he heard additional details about the Franklin party. Eskimos there told him that bodies had been found near the estuary of the Great Fish River. This information merely reinforced Rae's conviction that the bodies were part of Franklin's party. Furthermore, the Eskimos showed Rae more relics, including silver forks and spoons with crests of officers who had been on Franklin's ships. Other items included a plate with Franklin's name, a gold watch, a vest, and numerous other items. The Eskimos could not tell Rae what happened to the ships, but their most striking information was that the corpses showed signs of cannibalism. Rae decided to make haste to England immediately with the news.

In England his news was met with shock and disbelief. The relics he brought definitely proved that he had found news of Franklin, but no one believed that an Englishman would eat another Englishman. Everyone doubted Rae and thought he had been too gullible. Since they did not accept his story, the public made up its own explanation that involved massacre by the Eskimos. The alternative was simply too horrible to accept. Further, Rae was criticized for returning to England without going to the site where the bodies were found. This criticism led to the accusation that he had been eager to claim the £10,000 reward for finding evidence of Franklin. Public opinion ran high against Rae, the bearer of bad news, and against the Eskimos as murderous and barbaric.[5]

Lady Jane Franklin held Rae in great contempt for his unbelievable story, and his rough appearance and demeanor greatly offended her. She, like most of the public, disliked Rae for standing by the Eskimos' story. Rae maintained he was telling the truth, but Lady Jane retorted that he had believed lies told by "savages." She urged the Admiralty to delay the reward until further investigation by others. The Admiralty had doubts of their own about Rae's credibility, and he was forced to appeal many times for the prize money, which he eventually received and distributed a fifth to members of his party. Other Arctic explorers were knighted for their work, some of whom made less significant contributions to the map of the Arctic. Rae received the money, but no knighthood. The Admiralty and the public now began to lose interest in further searches for Franklin. Nevertheless, Lady Jane persisted in promoting additional searches, maintaining to the end that her husband had completed the Northwest Passage.

John Rae's accomplishments, surpassing all nineteenth-century Arctic explorers, were worthy of honors and international fame. No explorer ever approached Rae's prolific record: 1,776 miles surveyed of uncharted territory; 6,555 miles hiked on snowshoes; and 6,700 miles navigated in small boats. Yet he was denied fair recognition for his discoveries because he dared to utter the truth about the fate of Sir John Franklin and his crew. Author Ken McGoogan, in his biography of John Rae,[6] vividly narrates the astonishing adventures of Rae, who found the last link to the Northwest Passage and discovered the first evidence of Franklin's crew. A bitter smear campaign by Franklin's supporters denied Rae a knighthood and left him in ignominy for over 150 years.

19

Robert McClure Completes the Passage; Richard Collinson

Maps Coastlines, 1850

ROBERT MCCLURE'S EXPEDITION

In 1850 an expedition of two ships left England to search for Franklin from a westward approach. Commander Robert McClure (1807–1873) sailed the *Investigator*, and Captain Richard Collinson (1811–1883) commanded the *Enterprise*. Despite their separation early in the voyage and each becoming locked in ice, McClure and Collinson managed to survey some new coastlines and made important contributions to the map of the Arctic. By sledging about 200 miles from his icebound ship in Mercy Bay to Melville Island, McClure became the first explorer to tie together the east and west portions of the Northwest Passage and demonstrate that a water route, albeit frozen, existed through the North American Arctic.

The two ships became separated in the south Pacific, missing an intended rendezvous in Hawaii. Collinson waited in Oahu five days for McClure to catch up before proceeding toward Bering Strait. On the sixth day, McClure arrived in Hawaii and quickly departed trying to overtake Collinson. By taking a shorter route, McClure arrived at a supply ship rendezvous before Collinson, but refused to wait. McClure's haste and failure to travel into the Arctic with another ship broke what had become the common protocol of groups traveling in the Arctic. McClure explained his action as his interpretation of the instructions, but his ambition to be the first to complete the Passage or to find Franklin probably played a major role. In either case, there was big reward money at stake. As a result, McClure and Collinson each explored different routes and each eventually met with severe hardships alone.[1] McClure's route into the archipelago is shown in Chapter 15, figure 15.1.

McClure pushed ahead, helped by boats towing him around Point Barrow through gathering ice, and found the sea beyond clear of ice. He discovered and sailed into the Prince of Wales Strait that separates Banks Island from Victoria Island and came within 30 miles of open water separating Banks Island from Melville Island. There, heavy ice floes in Prince of Wales Strait stopped his progress. The *Investigator* wintered in the straits beginning in September 1850, held fast by ice until the following August.

In October 1850, McClure and the second master of the ship went on foot to the northeast entrance of Prince of Wales Strait and viewed Melville Island some 60 miles away. He knew the completion of the Northwest Passage was at hand, but ice had again won the day. In spring 1851, they made sledge trips along the coasts of Victoria Island and Banks Island looking for traces of Franklin's expedition and incidentally completing the coastal maps of those islands.

In the summer of 1851, McClure again attempted to complete the journey through Prince of Wales Strait, but as before, he met multiyear ice pack and had to retreat from the strait altogether. In September, before ice began to form again, McClure sailed the *Investigator* to the north side of Banks Island and took refuge in an inlet he named Bay of Mercy (now Mercy Bay), where they spent the winter of 1851–1852. In the spring, he sledged across the strait later named for him to Winter Harbor on Melville Island. This site was important as the place Perry reached on his famous voyage of 1820. There McClure left a message that later saved the lives of his entire crew, giving his position on Banks Island. Given the vastness of the Arctic, it is remarkable that messages left in cairns along the coasts were often found, and in some cases the messages were crucial to the success of an expedition.

In the summer of 1852, the ice in Bay of Mercy never cleared, and McClure's expedition faced another winter. The winter of 1852–1853 brought hard times to the *Investigator's* crew. Unlike the previous winter's activities of plays and music, this winter they faced greatly reduced food allowances, and there was a pervading dismal mood. Reduced rations in such a cold environment were insufficient to maintain both warmth and weight. The ship's surgeon wrote of the crew losing flesh, strength, and spirit. Soon scurvy and signs of starvation appeared, and two men went insane.[2]

McClure, perhaps slightly deranged himself, made plans to send half his crew on foot home to England; some he wanted to send toward Baffin Bay about 750 miles away, and others up the Mackenzie River, presumably to find rescue in some still occupied Hudson Bay Company post. If they should need to go all the way to Hudson Bay, the distance would be about 1,800 miles. He planned to send the weakest men, leaving the remaining men more likely to survive on the available food. His thinking was that he needed to keep the stronger men to sail the ship when it finally broke free of ice. Most of the men he wanted to send away were sick and could not have survived under the hardships of travel by sledge.

Fortunately for all concerned this plan, was never executed. The saving factor was the fleet sent out by the Admiralty in 1852 under the command of Captain Belcher. Their primary intent was to find Franklin, but part of their mission was to find McClure and Collinson. In September, the *Resolute*, under Captain Kellett, and the *Intrepid*, under Commander M'Clintock, arrived at Winter Harbor and left a cache of provisions for McClure and Collinson. Then they moved to a small island, Dealy Island, off the coast of Melville Island, and settled in for the winter. Eventually one of them found McClure's cairn giving information of his position and his discovery of the Northwest Passage. Naturally this new information caused great excitement for Kellett and M'Clintock, and they began making plans for reaching the men of the *Investigator* in the spring. Their plan included sending several sledge teams in directions that would intercept the *Investigator* if she broke free of the ice before they reached Bay of Mercy.

On March 10, 1853, they began the search, and one of the sledge teams led by Lieutenant Pim went toward the Bay of Mercy. This was early in the season for living out in the open, but they learned that given sufficient food, daylight, and good health, the terrain could be traveled. They made a very difficult 200-mile sledge journey over a rough and jagged ice surface toward the Bay of Mercy. One sledge became disabled in the process, and Lieutenant Pim took the remaining sledge with two men and covered the rest of the distance.

On April 6, Pim drew near the Bay of Mercy and seeing no ships or cairns assumed that McClure and the *Investigator* had already sailed. As he crossed the mouth of the bay on the ice, he noticed a dark spot farther up the bay, and his telescope revealed a ship. Pim went ahead alone to make better time and was a hundred yards from the ship before any of the crew noticed his presence. McClure and Lieutenant Haswell were out walking on the ice and greeted Lieutenant Pim with great enthusiasm—the visitor who seemed to drop from nowhere. Soon everyone who was able to crawl came on deck to see the rescuing stranger. When the sledge and two men arrived, the sailors aboard the *Investigator* gave out three loud and thankful cheers. When Pim and his companions saw the condition of the starving men and saw them eating their thin rations of weak cocoa and a piece of bread, they were moved to tears. They quickly brought out their own rations, including a large piece of bacon, and gave the crew the only substantial meal they had known for many months.

The next day, Pim and his crew, along with McClure and seven of his crew, set off on a twelve-day journey by sledge to meet Captain Kellett and the *Resolute*, still at Dealy Island. Two weeks later, additional crewmen from the *Investigator* arrived at

Dealy Island. The sickest of them rode on sledges, but even the relatively healthy men who pulled the sledges had to crawl part of the distance.

McClure, at this point, was faced with abandoning the *Investigator* permanently. Given his desire to finish sailing the Passage, McClure wanted to take his men back to his ship. Captain Kellett, the senior officer in this situation, was sympathetic to McClure's persuasion, but seeing the condition of the crewmen, he declared the ship must be abandoned unless the medical officers found the men physically fit and twenty or more volunteered to sail. As it happened, only four men volunteered, and McClure had to give up his ambition to be the first to sail through the Passage.

McClure returned to the *Investigator*, had it cleaned, secured the hatches, and left it in condition to sail should anyone find it in the future. Only now, six weeks after their rescue, did McClure finally allow the men to resume full rations. He had kept them on reduced rations in anticipation of continuing his voyage through the Northwest Passage and back to England. For reasons unclear, McClure ordered that all documents and diaries be left with the abandoned ship. Perhaps he wanted to be certain that only his version downplaying their desperate physical condition be published. Fortunately the Assistant Surgeon, Henry Piers, kept his journal. The senior surgeon, Dr. Alexander Armstrong, also retrieved his journal, and McClure allowed the interpreter, Johann Miertsching, to reconstruct his notes using the ship's log. When the abandoned *Investigator* was visited one year later, not one officer's journal was found. After the rest of the sick crew reached Dealy Island, Captain Kellett turned the *Intrepid*, also at Dealy Island, into a hospital ship where the emaciated and ailing men of the *Investigator* were nursed back to health.

In October 1854, Captain McClure faced a routine court martial for the loss of a ship, but was honorably acquitted and later knighted. Parliament awarded £10,000 to the officers and men of the *Investigator* for the discovery of the Northwest Passage, an honor that Lady Jane Franklin had promoted for her husband.

Narratives later written by Miertsching and Armstrong show Captain McClure as a somewhat unstable man driven by selfish ambition and given to using questionable methods to attain his goals. Despite these faults, they described McClure as a man worthy of the honors given him and gave him credit as a daring and skillful navigator.

In the summer of 2010, researchers from Parks Canada found the *Investigator* intact and sitting upright on the bottom of Mercy Bay in 36 feet of water where it had been abandoned in 1853. The masts and rigging had been sheared off by ice and wind, but the rest of the ship appeared to be in good condition. On the shore they found the contents of the ship where McClure had unloaded them, and three graves of men who died during their three-year ordeal.[3] The *Investigator* had a short career with the Royal Navy, purchased and converted from a merchant vessel in 1848 and abandoned in 1853.

RICHARD COLLINSON'S EXPEDITION

Captain Richard Collinson, sailing the *Enterprise*, became separated from McClure soon after rounding the Horn of South America, and despite two planned rendezvous, never connected with him again. Collinson arrived at the Arctic too late in the season to overtake McClure and turned back to winter in Hong Kong. He returned to the Arctic in the summer of 1851 and passed Point Barrow in late July, giving him an early start for exploring and mapping. Still, he repeatedly found ice blocking the way.

Collinson sailed into the Prince of Wales Strait and saw evidence that McClure had been there, but found it blocked by ice at the northeast end, just as McClure had. Bad luck persisted as Collinson sailed up the west side of Banks Island and again met an impenetrable ice barrier. Had this effort succeeded, Collinson probably would have found McClure's *Investigator* frozen into Mercy Bay. Failing in these attempts, Collinson found a wintering site in Walker Bay on Victoria Island near the entrance to Prince of Wales Strait. When spring arrived, he formed sledge parties to fan out in search for Franklin. Some of his search parties unfortunately wasted effort by unknowingly covering ground that McClure had searched the previous year. One search party came remarkably close to finding McClure, but thinking the sledge tracks they discovered belonged to Eskimos, they turned back to avoid an encounter. They had no weapons to defend themselves in case of an attack. Unfortunately, the tracks had been made by McClure's party. These near misses and overlapping searches reinforced the Admiralty's insistence on keeping ships together and coordinating their efforts. As a result, the *Enterprise* and the *Investigator* continued to be unaware of one another while separated by only 170 miles.[4]

In August 1852, the *Enterprise* broke free of the ice and sailed south along the coast of Victoria Land (now Victoria Island), gathering positional data for the uncharted coastline. Collinson entered the long inlet named Prince Albert Sound and discovered that Prince Albert Land and Wollaston Land attached to Victoria Land. He continued coasting and concluded that Victoria Land was an island that included other fragmentary coasts, Wollaston Land and Prince Albert Land. Now those names apply to peninsulas of Victoria Island.

Continuing along the south coast of Victoria Island, Collinson completed the map that had begun with work by Dease and Simpson. Of course, he had copies of the charts by Dease and Simpson and paid tribute to their high quality. At the end of the sailing season, he wintered in Cambridge Bay at the southeast corner of Victoria Island, only 160 miles from the remains of Franklin's ships. Collinson had the good fortune of a relatively ice-free season in the area, but deserves great credit for navigating the first large ship into these difficult waters. Others that previously saw these shores had traveled by canoes or small boats.

During the winter of 1852–1853 the crew of the *Enterprise* had friendly exchanges with visiting Eskimos and asked them to draw their mental maps of coastlines for the area. The result was inconsistent with maps the mariners had made, but they seemed to show a ship some distance away. The only interpreter, Johann Miertsching, was traveling with McClure, so clarification was not possible. Collinson remained skeptical that the Eskimos' mark indicated ships, but later, pieces of wood identifiable as part of Franklin's ship were found on an island in nearby Dease Strait.

In the summer of 1853, Collinson, with three sledges, twenty-four men, and nine dogs, proceeded along the north shore and discovered cairns left by Dr. John Rae, who had come there looking for Franklin in 1851. Once again, Collinson found himself searching in areas that had already been covered. As we have seen, Rae had not found Franklin but was the first to hear from the Eskimos that British sailors had been lost in the area. At the time, Rae, and now Collinson, were but a short distance from the site of the missing *Erebus* and *Terror*.

Collinson decided to push ahead to cover new land that Rae had not searched. In doing this, he contributed another segment of coastline to the emerging map of the Arctic. Unfortunately, Collinson, like Rae, traversed the west side of Victoria Strait near the shore of Victoria Island. Had he chosen the east side of the strait he would have found the site where Franklin's party had abandoned their ships. In returning to Bering Strait, Collinson and the *Enterprise* spent another winter (1853–1854) along the north coast of Alaska.

In the summer of 1854, *Enterprise* made rendezvous with the supply ship, *Plover*, which had been waiting since 1852 for Collinson to return. Collinson and company then returned to England via the Cape of Good Hope. His return was met with little enthusiasm. The Admiralty and the public had lost hope of finding Franklin. Collinson's voyage had yielded no news of Franklin, but he had mapped new coastlines and determined the extent of Victoria Island and its peninsulas. After his retirement from the Royal Navy, Collinson was knighted and served as an editor, writing an extensive introduction for the Hakluyt Society's book on Martin Frobisher.[5]

ICE MOVEMENTS IN THE ARCTIC

The search for a Northwest Passage in the nineteenth century, during the waning years of the Little Ice Age, might be considered tragically unlucky timing. Men and ships were entrapped by the ice, and in several cases the explorers abandoned their ships, and men died. Ice core data supports the notion that the nineteenth century was the latter part of the Little Ice Age cold period. Was the effort really a failure? Countless volumes of books and articles on nineteenth-century exploration give detailed accounts and analyses of the expeditions, but the volumes of scientific and meteorological observations collected during those expeditions have been largely unexamined.

Arctic researchers K. Wood and J. E. Overland compiled temperature and ice data from forty-four reports and narratives of sailing expeditions, some with auxiliary steam, spanning years from 1818 to 1910, and compared them with data from 1971 to 2000.[6] Most of the nineteenth-century data examined came from expeditions that wintered in the Arctic and followed the naval practice of collecting scientific data. The naval officers on those expeditions faithfully recorded air temperatures, ice thickness, and dates of freeze and thaw. Overland and Wood point out that the lowest temperatures may not have been recorded accurately due to the use of mercury thermometers that froze at -38°F.

Wood and Overland found that "the extent of summer sea ice during the nineteenth century, insofar as it is shown in patterns of navigability inferred from ship tracks, the direct observations of explorers, and a number of native accounts, is remarkably similar to present ice climatology." Present refers to the thirty years from 1971 to 2000.

A look at routes followed by Arctic exploration expeditions in the nineteenth century shows that most of the Northwest Passage was navigated with gaps as short as 90 miles between unconnected points. Despite being called a failed effort, the route was almost completed by several expeditions. Wood and Overland's ice frequency map suggests that a sailing vessel in the late twentieth century would have had almost the same likelihood of encountering sea ice blocking the route as did those in the nineteenth century. See figure 19.1.

Wood and Overland conclude that the ice and climate observations made by the nineteenth-century explorers are not consistent with the severe conditions associated with the concept of a Little Ice Age derived from ice core data. From year to year, the nineteenth-century explorers experienced great variations in ice conditions. Such variation accounted for the exceptional luck met by Sir William Parry, allowing him to sail all the way to Melville Island in 1820. Neither he nor others were able to duplicate that feat for years. Also, annual variability explains Franklin's voyage through Peel Sound in 1846, while all the search parties in later voyages met solid ice on that route.

William Parry became a favorite of the Admiralty partly because of his lucky break in ice conditions. He later realized the role of ice variability when he repeatedly met misfortune on other voyages. The Admiralty made plans for expeditions with the expectation of good ice conditions, overlooking that such conditions were inconsistent. They ultimately felt defeated by the ice, but Wood and Overland show us that the outcome might have been the same for similar vessels in the late twentieth

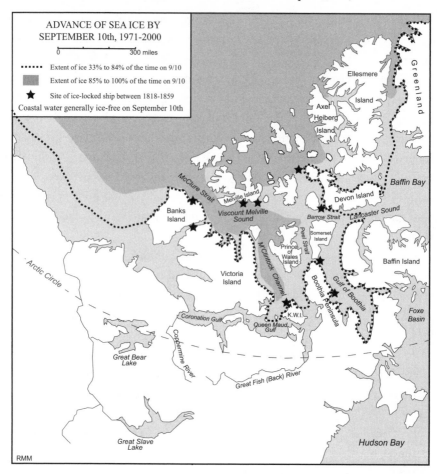

FIGURE 19.1 September tenth is taken as the usual time of the minimum extent of ice. The darkest shade shows a higher frequency of ice movement into McClure Strait and southeast into M'Clintock Channel. Both McClure, on the north coast of Banks Island, and Franklin, at King William Island (KWI), were caught in this area of the year's earliest probable presence of sea ice. Map was derived from the Overland and Wood article.

century. Despite their sense of failure, those early explorers found all of the possible routes that make up the Northwest Passage, and mapped almost the entire coastline of Arctic North America. They came within a hundred miles of sailing through the Passage.

Another question related to the success of sailing through the Arctic Archipelago concerns the movement of ice. Certain waterways among the islands become ice-bound earlier and more frequently than others. This variation results from movement of ice into the archipelago from adjacent oceans.

Researchers T. Agnew, A. Lambe, and D. Long found evidence that multiyear ice forms in the western and northern regions of the Canadian Archipelago and slowly migrates southeast.[7] In the early winter months, sea ice from open water of the Arctic Ocean advances into the Archipelago at the rate of 6 to 9 miles per day. During this time, ice in the channels is unconsolidated and mobile. In September and October (the months when sailing came to a halt), southeastward movement of ice passes through McClure Strait toward Barrow Strait at the rate of about 2 1/2 miles per day. Also, ice moves into M'Clintock Channel during this period at the same rate. This movement into M'Clintock Channel is especially consistent from year to year. Also M'Clintock Channel appears to be receiving ice moving south from channels in the Queen Elizabeth Island group to the north. The Amundsen Gulf and Lancaster Sound are ice-covered by October in about three out of five years, i.e., about a 60 percent probability of ice in those waters by the end of October. Ice may also enter the Archipelago via Lancaster Sound from Baffin Bay.

By late winter (March), the ice in the central part of the part of the Archipelago has thoroughly consolidated and become firmly attached to land with little possibility for movement. However, ice may still be moving into Lancaster Sound from Baffin Bay to the east. Within Baffin Bay, ice drifts counterclockwise with the westward component pushing into Lancaster Sound at about 9 miles per day. Understanding the seasonal movements of ice through the Canadian Archipelago helps one realize why the Northwest Passage presented such a challenge to the nineteenth-century mariners.

20

Elisha K. Kane Barely Survives, but Maps New Land, 1853

AN AMERICAN, ELISHA Kent Kane (1820–1857), undertook a search for Franklin in 1853. He was sponsored by Henry Grinnell, the American financier and backer of Lady Jane Franklin's search expeditions. Kane's quest for Franklin headed north along the west coast of Greenland—a direction not even mentioned as an alternative in Franklin's orders. It is peculiar that so many searchers expected to find Franklin in such unlikely places. It would have been most unusual for a loyal and trusted Royal Navy officer to make such a diversion from his orders. In that sense, Kane's expedition was pointless from the beginning. Kane, however, had a second objective. He hoped to reach the still mythical ice-free polar sea and sail to the North Pole.

Born in Philadelphia in 1820, Kane graduated from the University of Pennsylvania medical school in 1842. The next year he joined the U. S. Navy as Assistant Surgeon. In 1851 he received an appointment as senior medical officer on the *Advance* in the first Grinnell Expedition to the Arctic in search of Sir John Franklin. That expedition, though unsuccessful in finding Franklin, had joined the British search for Franklin at Beechey Island, known to be Franklin's first winter encampment. They conducted the search up the Wellington Channel where they became locked in ice for the winter. Their companion ship, the *Rescue*, had to be abandoned because of severe ice damage. At the first opportunity, the *Advance* returned to the United States.

In 1853, Kane himself organized and commanded a voyage named the Second Grinnell Expedition—again aboard the *Advance*. From the beginning, many obstacles hampered this venture. Though he had been in the American Navy for ten years and had spent a short time in the Arctic as ship's physician, Kane had never commanded a ship and had spent much of his duty on shore. He knew very little about navigation and had no experience leading men. His inexperienced crew included some unsavory characters—two of whom, Godfrey and Blake, kept the ship in a near constant state of quarreling—and Kane was unable to maintain proper discipline. He would put the troublesome two in confinement briefly but could not spare them for long as he needed every hand available. The crew came to hate Kane.[1]

Kane sailed the *Advance* north through the Baffin Bay ice and into new territory in the water body that now bears his name, Kane Basin. He had gone farther north than any other explorer, but it was August, time for them to be finding winter quarters. The crew argued for turning back, but Kane felt they should continue. Eventually he accepted his officers' advice to stop for the winter. They found a suitable bay, which he named Rensselaer Bay for his family's estate (figure 20.1).

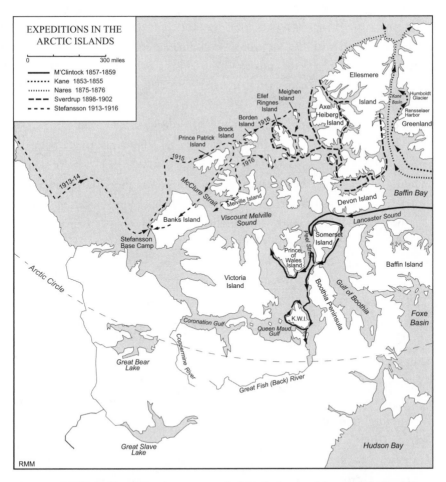

FIGURE 20.1 M'Clintock's and Kane's voyages marked the final stage of the search for Franklin. Nares aimed primarily for the North Pole; Sverdrup and Stefansson simply went to map new lands.

Kane's inexperience with planning and managing an expedition became serious at this time. His crew, most of whom had never spent a winter in the Arctic, experienced the oppressive effect of the permanent night in the polar landscape. Not only were they plagued by the darkness; they also had to eat their food cold because Kane had not brought enough fuel. By the end of February, they were almost out of coal and oil. Water could not be melted for washing or for tea; they could not make fresh bread. The galley stove became useless. Everyone became disagreeable, and Kane could not keep the crew in order. Also, scurvy plagued some of the men, and an illness among the dogs killed all but six of the original fifty. The lack of dogs now required that Kane's search for the open polar sea the following spring must be done by manhauling the sledges.

Kane began preparations for the sledge journey even though winter winds with deadly temperatures were still piling snow into drifts. On March 19 he sent eight men and a sledge to establish a provisions depot along the route they planned to take for the push toward the supposed polar sea. The temperature was -40°F, blowing snow hit their faces with the force of sand grains, the ice surface was rugged with pressure ridges, and the sledge was too heavy and bulky to move over the ridges. After some days, the sledge men decided to turn back because they had been unable to establish the depot. Four of the men were so badly frostbitten they could not walk, and they soon collapsed. Immediately they realized they were all in serious danger. Three men made a dash for the *Advance*, 30 miles away, to get help. They marched without taking time for food or sleep and made the trip to the ship in thirteen hours. The fact that they averaged 2.3 miles per hour under those harsh conditions is testimony to their determination. Their effort left them staggering, delirious, badly frostbitten, and near collapse from exhaustion. Only one, Ohlsen, could speak enough to explain the situation to Kane.

Kane took seven men and began immediately on the rescue trip. Poor Ohlsen, tied onto a sledge, had to return with them as a guide. Again the ridged and jagged ice surface on the way soon made the heavy sledge impossible to manage, and after 20 miles they resorted to walking. Two men had to support Ohlsen, who by then was only half conscious. After another four hours, they found the tent and the missing men who were barely alive. The rescue party, themselves exhausted, needed rest before starting the return journey. Unfortunately, the tent was too small for a party of twelve men, and they had to sleep in two-hour shifts while others stood outside stamping their feet to keep from freezing. Then, with the rescued men strapped to their sledge, they set off for the *Advance*.

The journey back to the ship was grueling. Ice ridges, formed by the compression of ice floes, confronted the men with long, jagged barriers 10 to 20 feet high, requiring them to chop pathways over the ice. At each ridge of ice, the sledge had to be unloaded and carried over the barrier while the sick lay on the ice waiting to be helped across. After 10 miles, the healthy men began to fall from fatigue. Only Kane and Godfrey could still do the work. At this point, Kane and Godfrey decided to push ahead 9 miles on their own to retrieve the sledge left behind earlier. Godfrey had been one of the most contentious men on the crew, but in this dire situation he certainly proved his worth. Before the two had gone the 9 miles, Kane became delirious and began to stumble and fall. Godfrey helped him along until they reached the sledge. When the remainder of the party caught up, everyone rested before continuing.

This journey continued with men raging like maniacs one moment and babbling like children the next. Godfrey said later that he had contemplated suicide for the first time in his life. The trip ended at the ship with all hands alive, although two died later—one from complications of amputations due to frostbite, the other from tetanus. Kane felt the anguish of failure enveloping his expedition. He had not found Franklin or the ice-free polar sea, and the crew was thoroughly debilitated. Unfortunately, Kane's troubles were by no means ended. The remaining food consisted of salt meat, the fuel was reduced to 750 pounds of coal, and the still fractious crew comprised many sickly men. The ship had been stripped of wood for burning through the previous winter, and removing additional wood would make her unseaworthy.

Following this harrowing winter of 1853–1854, Kane's fortunes took a turn for the better. He had a chance to visit an immense glacier seen by some of his crew on a scouting trip the previous summer, which he had already named the Humboldt Glacier in honor of the world's most highly respected scientist at the time. When he reached the sheer face of the glacier where it met the sea, it was so steep that he was unable to climb it to see the top. Kane and one other man climbed a nearby cliff to have a look at the glacier.

This massive glacier still holds the record as the world's largest and has its origin in the vast Greenland Ice Cap. Kane's report on this 60-mile-wide glacier proved to be significant to the controversy on continental glaciation initiated by Louis Agassiz, the Swiss naturalist.[2] Agassiz developed his theory as an explanation for deposits of erratic boulders found in northern Europe and North America, but up to that time, no person had seen a glacier of such continental proportions. Many skeptics could not imagine ice covering an area greater than the alpine glaciers seen in the Swiss Alps. Kane's descriptions of the glacier showed that he was familiar with glacial studies. He understood that a glacier is a massive flow of ice. He wrote, "Repose was not the characteristic of this seemingly solid mass; every feature indicated activity, energy, movement. A more impressive illustration of the forces of nature can hardly be conceived."[3] Kane's observations of the Humboldt Glacier were credited with influencing opinions of geologists on previously unanswered questions concerning continental glaciation.

Another positive development came when a scouting party returned with the news that Kane Basin had a 30-mile-wide channel exiting to the north. Kane quickly named it Kennedy Channel after a friend. He was able to sail the extent of the Kennedy Channel as far as a promontory he named Cape Constitution at 80°27′ N, a new "farthest north" point at the time. This, along with an extensive survey and resultant map of the Kane Basin coastline, reassured Kane of his expedition's contribution.

There was one more day of excitement when one of his party reported climbing a cliff and seeing an endless vista of open water shining in the distance. Kane now knew he had found the open polar sea that had been hypothesized for centuries. He was sure that his expedition would be famous. Unfortunately, there was no open water. It was no doubt a mirage—perhaps ice reflecting the sky in the distance—combined with wishful thinking. Kane's expedition is remembered, of course, but not for finding an ice-free polar sea.

In August 1854, the *Advance* had been in Rensselaer Bay for a year, locked in ice with no chance of getting her free. They all faced the prospect of spending another winter in the ice. Kane still held hopes of finding Franklin, who had been dead for seven years in a place about 800 miles to the southwest. Kane was again discouraged, as were the crew. Kane became aware that some of the crew were having secret meetings, and he saw groups whispering among themselves. After some time, one of the men, William Morton, told Kane that a group of them planned to make a run for the nearest civilized settlement in Greenland, Upernavik, which was 700 miles to the south. This news set Kane raging against the men as dishonorable traitors. It also left him little choice but to give his approval. If he insisted on their staying, assuming that was even a viable option, the crew's morale would have become worse. He spoke to the men, telling them of the extreme dangers they would face on such a long journey in a weakened condition and with minimal food. He gave them twenty-four hours to make their final choice. The answer shocked Kane. Most of the crew, including four of his officers, wanted to make the trek. Kane persuaded some to change their minds, but in the end, eight of the crewmen wanted to leave. This left nine men on the ship.

Kane made the defectors sign a statement acknowledging that when they left, their connection to the expedition would be terminated. After three days of organizing, the defecting group prepared to go. Kane, though seething with anger, behaved cordially toward them and even had a round of champagne toasts. He was certain, however, they would have insurmountable difficulties and would want to return. A few days after the defectors departed, two of them, Blake and Riley, returned, having fallen through the ice and nearly drowned. But Blake then left again to rejoin the defectors. The ship's company now stood at ten, and those men settled in for another long winter. They hoped to barter for seal meat with the Eskimos, but the Eskimos did not come around. Kane himself took advantage of a rat infestation on the ship and managed a diet of fresh rat meat that other members of the crew refused to eat. As a result, Kane made it through the winter in good health, while half the crew suffered from scurvy.

In December, two of the defectors returned to the ship, helped along by Eskimos. They reported that the remaining defectors were in a stone hut. In fourteen weeks they had managed to travel only 150 miles of their 700-mile journey. Now they begged help for themselves and their comrades. Kane responded immediately with one hundred pounds of provisions, which he persuaded the Eskimos to carry back to the starving men. He could not be sure the Eskimos would take the food to the men, but he had no choice. The Eskimos delivered the food, but the stranded men were now dependent on them for additional food. Eskimos brought the hungry men only small amounts of food at a time, bartering for trade goods each time. In this situation, the men had little chance of accumulating enough food to continue their journey. When they tried to barter with the Eskimos for a sledge and dogs to make another trip to the ship, the Eskimos, wanting to keep the men dependent, turned hostile.

After finally escaping from the Eskimos, all the men made the trek back to the *Advance*. Kane took them in, gave them medical attention and food, but kept them segregated from the rest of the crew. Their return put a hardship on everyone. Now there were seventeen mouths to feed again. The returning men had lost all their food and equipment—everything but the clothes they wore. Kane decided to share provisions, but not to treat them as equals with the rest of the crew. He made two groups—the faithful and the unfaithful.

In May 1855, after a long and contentious winter, Kane realized they must abandon the *Advance*. They were in no condition to wait through the summer to see if the bay would open, and he was no longer sure the ship had enough wood to be seaworthy even if it did break free of the ice. They now needed to move three boats, four disabled men, and 1,500 pounds of supplies to open water 80 miles south. They shuttled the loads a little at a time, moving one boat forward a mile, and then returning for another load. This constant moving back and forth caused each man to travel 316 miles to cover the 80-mile distance. Kane alone, by making extra trips back to the ship, traveled a total of 1,100 miles by the time they reached open water at Naviliak on June 18 after thirty-one days of tortuous travel.[4] They were not yet safe. They could, however, travel by boat the rest of the way to Upernavik. Travel through icy waters held terrifying hazards, and they were near starvation before finally reaching safety on August 6, 1855. There they learned that others had discovered Franklin's fate.

The Second Grinnell Expedition, led by Kane, had charted some 960 miles of coastline, discovered the Kane Basin, Kennedy Channel, and the Humboldt Glacier—altogether a significant contribution to the mapping and science of the Arctic. Given the overwhelming problems they faced, it is amazing that they lost only three men. Kane received medals from Congress, the Royal Geographical Society, and the Société de Géographie honoring his achievements. A lunar crater is named for Kane, and in 1986 a United States postage stamp was issued honoring Elisha Kent Kane. His health never recovered from the struggles of his last expedition, although he went to the American Naval Hospital in Cuba to convalesce. He died in 1857. His funeral train from New Orleans to Philadelphia was, except for Lincoln's, the longest such train of the nineteenth century.

21

Francis L. M'Clintock Extends the Map and Learns

What Happened, 1857

LADY JANE FRANKLIN'S intuition was right. She maintained that the search for Franklin should concentrate south of Lancaster Sound. The Admiralty continued to send expeditions to the north and west, even though they were unlikely to find Franklin by repeating the same mistake. It made no difference to her that on January 19, 1854, the British government officially pronounced the men of the expedition dead. She was determined to find them, dead or alive, in part to vindicate her husband as the one who completed the Northwest Passage.

Jane Franklin had spent much of her personal money on three unsuccessful expeditions searching for her husband. She was frustrated that ships had been sent in what she thought was the wrong direction. Further, she felt that McClure had been given credit prematurely for completing the Northwest Passage in 1853. Since no documents had been found regarding Franklin's voyage, he might have made the link before McClure. She had a strong intuition that King William Island must be searched. The area of previous searches had been determined in part by the location of ice-free waters. If a channel was blocked by ice, they thought it had also been blocked for Franklin. As yet, they were not fully aware of the variation in ice distribution during an occasional warmer year.

In 1857, Lady Jane arranged for Francis Leopold M'Clintock (1819–1907) to sail the *Fox*, a ship she financed with £2,000 of her own money plus public subscription. M'Clintock's narrative lists some 240 donors making contributions totaling £2,981, with amounts ranging from 2 shillings and 6 pence to £500 from "a commander in the merchant service."[1] The Geographical Society of Paris contributed £40. Many donations were made to collection boxes by persons soliciting a few coins. This wide variety of people giving money reflected the great interest the public had for Franklin. In-kind donations included equipment for the ship, preserved potatoes, and curiously, twenty dozen containers of "Isle of Wight sauce" from a Mr. Tucker of Newport.

Captain Collinson, who had earlier participated in a search for Franklin, gave £20 and also became Lady Jane's business manager for the expedition. Public interest in the lost explorers led to financial support but also produced some oddities. In one instance, a seance held in 1849 channeled the spirit of a 4-year-old who told a member of her family that Franklin's ships could be found in Victoria Strait. This turned out to be correct.[2]

M'Clintock brought just the right experience to this expedition. A man of the same thinking as John Rae, M'Clintock was expert in sledging with either dogs or manpower. Like Rae, he understood that travel and survival in the Arctic had to be done in the ways used by Eskimos—wearing the same clothing, eating the same food, and making the same shelter. Adaptation to the environment was the key. The inflexible use of English woolen clothing, tents, bedding, and tinned food required massive loads for sledges.

M'Clintock also had extensive experience as a surveyor and had mapped Melville Island, Prince of Wales Island, and Prince Patrick Island. As an officer in Captain Kellett's expedition of 1852–1854 in the *Herald*, M'Clintock had traveled 1,400 miles by sledge and mapped 800 miles of new coastline. Best among all his qualifications, M'Clintock agreed with Lady Jane Franklin and was eager to make the search. Lady Jane had the support of powerful men both in the navy and Parliament. In addition, her iron determination to vindicate her husband prevailed when the Admiralty's will began to flag.

The *Fox* was a steam-sail yacht of 177 tons. Although the small size of the *Fox* limited the size of the expedition, it had advantages in maneuverability. M'Clintock had no assignment with the navy at the time, so he was free and willing to command the expedition to King William Island. The expedition was meagerly funded, but M'Clintock saw that handicap as part of the challenge. Extensive modifications were necessary to make the *Fox* seaworthy in the Arctic—strengthening sides, cross beams, prow, and screw. M'Clintock chose twenty-five men experienced in both sailing and driving dog sledges in the Arctic. Most of them had been on previous searches for Franklin. M'Clintock and the other officers of the *Fox* declined to accept any money from Lady Jane for themselves. She also wrote an agreement to free M'Clintock of any liability during the expedition and to reward him with the *Fox* at the end. M'Clintock declined to accept the *Fox*.

At their departure on the last day of June 1857, Lady Jane arrived to wish them well, and the crew gave her a hearty cheer as she left the deck of the yacht. M'Clintock read her instructions to the crew as they got under sail. First, they should rescue survivors; second, find the documents of the expedition, and third; confirm that the Franklin expedition had made the discovery of the Northwest Passage. This voyage commenced twelve years after Franklin's, and the possibility of finding survivors was slim.

Unfortunately, in 1857, ice conditions in Baffin Bay were bad, and the *Fox* became locked in ice for the winter far from Lancaster Sound. By April they had drifted with the floe for 250 days and traveled 1,385 miles southward. Now the floe began to break up with dangerously high seas, and the little *Fox* was knocked around as floes pushed back and forth in surges strong enough to make men fall to the deck. Any one of these shoves could have crushed the *Fox*. M'Clintock's greatest concern was for possible damage to the rudder or the screw. After several days of rough treatment, the ship was suddenly free of ice and sailing in open water again.

Now M'Clintock had to retrace his voyage of the previous summer up the west coast of Greenland and across Baffin Bay north of the persistent ice pack. Even this crossing was not without danger. Concerning the crossing of Baffin Bay, M'Clintock wrote on June 11, 1858, "how closely danger besets the Arctic cruiser, yet how insidiously; everything looks so bright, so calm, so still, that it requires positive experience to convince one that ice only a very few inches, perhaps only three or four inches above water, perfectly level, and moving extremely slow, could possibly endanger a strong vessel!"[3] Finally, in July they reached Lancaster Sound and proceeded westward, still among ice floes retarding their progress (figure 20.1).

At Beechey Island, M'Clintock found provisions and coal in good condition, left from Belcher's expedition of 1852–1854. This was a most welcome find as the *Fox* had used a great amount of coal pushing through the ice in Baffin Bay, and the extra time had reduced provisions. On August 18, the *Fox* sailed down Peel Sound on the same route taken by Franklin twelve years earlier, but they soon encountered ice from shore to shore. They immediately turned back and headed for Prince Regent Inlet hoping to pass through Bellot Strait back into Peel Sound. This move was highly risky, especially in a year of little open water. Bellot Strait is narrow and was unlikely to be ice-free. Furthermore, they were still uncertain at that time if Bellot Strait actually existed.

Bellot Strait is similar to a fjord. Between cliffs 1,500 feet high, the strait is 20 miles long and barely 1 mile wide at the narrowest point, which is near the east entrance. Ice from Peel Sound had pushed westward into Bellot Strait, choking the opening at the narrow east end. For several days, M'Clintock attempted to navigate into openings in the ice pack and was repeatedly stopped. He progressed far enough to see that Peel Sound beyond was still full of ice. This futile effort lasted until September 12 when M'Clintock decided to find harbor for the winter of 1857–1858. In the winter, M'Clintock told of opening a cask of sea biscuits and finding a live mouse inside. The cask was watertight and had no holes, so they concluded that the mouse had entered the cask before the biscuits were packed the previous June.

M'Clintock concluded that the next stage of his search would be by dog sledge, and he began immediately preparing for a trek to the mouth of the Great Fish River; on the return, he would examine the shore of King William Island. He planned three separate sledge parties, including an experienced dog sledge driver with each. The plan involved one party to explore the west coast of Boothia, one to travel to Prince of Wales Land (now Island), and the third, led by M'Clintock himself, to sledge to the Great Fish River. The search in nearby areas could begin immediately. The remainder would begin the following March while seas were still frozen. M'Clintock's objectives were to complete the search for Franklin and the mapping of Arctic coasts in one of the few remaining unmapped areas of the Canadian Archipelago. On October 19, the first of the sledge parties left the *Fox* exploring southward along the west coast of Boothia. They returned in good health after nineteen days, though suffering from exposure, with no news of Franklin. The rest of the winter required bracing against the cold, monotony, and the confinement that comes with -40°F temperatures and howling winds.

M'Clintock and his sledge party set off for the Great Fish River in mid-February, a month earlier than planned, with temperatures still at -30°F. He expressed some misgiving about leaving so early in the year. The weather had been extreme, and they expected to be out for three weeks. The equipment his party took included a small tent, felt robes, sleeping bags with double blanketing, and fur boots for sleeping. Usually they made a snow hut at the end of each day, which provided much better shelter than a tent. The only extra clothing was additional foot gear to assure dry feet each day.

On the west coast of Boothia on March 1, they met four Eskimos, one of whom had a button from a naval uniform. The Eskimos told them of starved men on the nearby island, although they had not actually seen the men. The next day, an entire village of forty-five Eskimos arrived with relics from the Franklin expedition to trade with M'Clintock. None had seen the starved men alive, but a few had seen bones on the island. One of the Eskimos told of a three-masted ship being crushed by ice in the sea west of the next island, which was King William Island. He said the ship sank out of sight, but that all the people on the ship had landed safely.

On April 2, M'Clintock and a party of twelve men with a combination of man-pulled sledges and dog sledges set off for King William Island. As they traveled, they met Eskimos who told them there had been two ships offshore. One had sunk in deep water and the other had been forced on shore and badly broken up by the shoving ice. The Eskimos had entered the ship to retrieve items of value to them and found one dead man of very large size. Further, they said that all the remaining white people had gone away to the "large river," which M'Clintock took to mean the Great Fish River, and they had taken boats and supplies with them. M'Clintock understood that this event had occurred in the autumn several years earlier, and that during the following winter the Eskimos had found bones of the men near the mouth of the river.

At this point, M'Clintock split the sledge party into two parts. One party went directly toward the west coast of King William Island to find the site of the ships, while M'Clintock led the other group to the south coast of King William Island and to the mouth of the Back River. As M'Clintock reached the south shore of King William Island, he encountered more Eskimos who had six pieces of silver table service with the crest of Franklin and three of his officers, which he bought for the price of four needles each. They heard from the Eskimos that books and papers on the ship had been destroyed by the weather. After buying all the relics available, M'Clintock also purchased seal's meat, dried salmon, and frozen blubber, a high-energy food that he had learned to like. From one informant, M'Clintock learned that many of the white men had dropped dead on their way to the Back River. Some had been buried, and others were still lying on the surface. None of the Eskimos had actually witnessed the men dying, but had come upon their bodies the following winter. M'Clintock reached the estuary of the Back River in mid-May amid winter storms. He traversed the shores of the estuary but found no trace of the missing men.

Returning to King William Island and sledging westward, M'Clintock found a cairn which they dismantled, stone by stone, looking for messages or documents, but found nothing. They even dug the frozen ground below the cairn with pick axes, but to no avail. It was strange that this second cairn left by the Franklin expedition, like the first on Beechey Island, contained no information. M'Clintock now knew that they were moving toward the wrecked ships along the route of Franklin's men as they tried to reach a safe haven. Knowing this, they kept a sharp eye for any sign of those men that might be seen through the snow cover. On May 24, 1858, M'Clintock came upon the first evidence of Franklin's expedition—a partially exposed human skeleton. The type of clothing and the neckerchief knot on the skeleton showed the man had been a steward or officer's servant. The face-down position of the skeleton suggested that the young man had simply dropped dead as he walked.

Later M'Clintock found a recently made cairn left by his Lieutenant Hobson, leader of the party that had gone directly toward the site of the ships. The message said that they had seen no trace of the boats but had found a written record at Point Victory on the northwest coast of King William Island. The record was a printed form with blanks for filling in position and date. The note revealed the following information:

28 May 1847. H. M. ships Erebus and Terror wintered in the ice in lat. 70°05′ N.; long. 98°23′ W. Having wintered in 1846–7 at Beechey Island, in lat. 74°43′28″N., long. 91°39′15″ W., after having ascended Wellington Channel to lat. 77°, and returned to the west side of Cornwallis Island.

Sir John Franklin commanding the expedition. All well.

Party consisting of 2 officers and 6 men left the ships on Monday 24th May, 1847.

Gm Gore, Lieut., Chas. F. Des Vaeux, Mate.[4]

M'Clintock immediately caught the error in the time the expedition had wintered at Beechey Island. Franklin had actually been to Beechey Island for the winter of 1845–1846. Otherwise, this note told of a successful expedition up to May 1847. M'Clintock now realized that by the autumn of 1846, after leaving Beechey Island, Franklin had reached a point about 12 miles off the northwest coast of King William Island when he was stopped by ice. There his expedition had spent the winter of 1846–1847 and up to the time the note was written in May 1847. The terse notation, "All well," suggests that no serious loss of life had occurred up to that time, and the crew had no anticipation of difficulties to come. The eight men who left were probably a survey party sent to explore and map the extent of King William Island, which was unknown at that time. Almost a year after the first message, Captains Crozier and Fitzjames had later opened the same cairn to add some updated information to the original note. Their handwritten entries in the margins told the tragic story.

April 25, 1848—H. M. ships, Terror and Erebus were deserted on the 22nd April 5 leagues N.N.W. of this, having been beset since 12th September, 1846. The officers and crews consisting of 105 souls, under the command of F. R. M. Crozier, landed here. Sir John Franklin died on the 11th June, 1847; and the total loss by deaths in the expedition has been to this date 9 officers and 15 men.

F. R. M. Crozier   James Fitzjames
Captain and Senior Officer   Captain H. M. S. Erebus.
and start tomorrow, 26th, for Back's Fish River.[5]

In M'Clintock's words, "A sad tale was never told in fewer words." In the eleven months from May 28, 1847, to April 25, 1848, everything had changed for the Franklin expedition. Franklin himself had died only two weeks after the first note written by Graham Gore. If Franklin was sick at the time, his illness apparently was not yet considered serious enough to mention, but very soon he died. His grave—probably not far on shore from the ships and probably not very deep, given the difficulty of digging in the frost—has never been found. His expedition reached within 100 miles of the Queen Maud Gulf before being blocked by ice. They had sailed through 500 miles of unexplored territory and must have felt confident of completing the Northwest Passage once ice thawed the following spring.

One year later, in April 1848, Captain Crozier, who took command when Franklin died, had abandoned the expedition and tried to take his 105 starving men to a place of refuge. The nearest destination would have been the Hudson Bay Company post on the Great Slave Lake, more than 1,000 miles away. The company of the *Erebus* and *Terror* must have been on reduced rations for much of the winter of 1847–1848, and the men wasted enough to be in a weakened condition before they left the ships. The ships left England with sufficient provisions to last until the summer of 1848. By that time, they had expected to meet the supply ship waiting near Bering Strait.

M'Clintock saw that his own provisions were low, and he made preparations to return to the *Fox*. Traveling along the west coast of King William Island, he found other relics, including a 28-foot boat mounted on a sledge that the escaping men had brought only a short distance. A large quantity of tattered clothing lay in the boat, along with portions of two human skeletons. The boat weighed about eight hundred pounds, and mounted on the heavy sledge, it made a 1,400-pound dead weight for the weakened men to pull. The crews must have thought they would need the boat to leave King William Island and reach the mainland. More likely, however, they could have crossed the 2-mile wide Simpson Strait on ice and reached the mainland with no boat. Furthermore, traveling with such an incredible weight went against all that had been learned up to that time about travel in the Arctic. Survival depended on traveling light. Besides the provisions they could muster, they carried almost anything of value, including silver table service and an "amazing quantity of clothing"—most of which was of no use to their survival.

Other items found in the boat included five watches and two double-barreled guns—one barrel in each of them was loaded and cocked. Several small books were found, including *The Vicar of Wakefield* and scriptural or devotional readings. A brief listing of the items M'Clintock found serves to illustrate their seeming lack of awareness of the dire situation they faced. He found different types of boots, such as sea boots, cloth winter boots, heavy ankle boots, and strong shoes, silk handkerchiefs, towels, soap, tooth brushes, and combs. Other items included twine, nails, saws, files, bristles, wax ends, sailmakers' palms, powder, bullets, shot, cartridges, knives (including dinner knives), needle and thread cases, and two rolls of sheet lead. The guns, bullets, powder, and hunting knives would have been essential if they found game. M'Clintock found a bit of tea and forty pounds of chocolate, but no biscuits or meat. Also in the boat were eleven large silver spoons, four silver tea-spoons, and eleven silver forks, all bearing crests of Franklin or one of the other officers, but no iron spoons of the type issued to seamen.

M'Clintock noticed that of the twenty or thirty men assigned to this boat, only two bodies remained with no sign of graves nearby. Apparently, desperation forced the other men to abandon the heavily loaded sledge and continue only with what they could carry. Another oddity he noticed was that the boat and sledge were pointed northeast—back toward the direction of the ships—and the survivors would have been headed the opposite direction. M'Clintock deduced that the party traveling with this boat decided to return to the ship for a fresh stock of provisions, and these two dead men were unable to keep up. They must have been left in the boat until the others could return with food for them. M'Clintock surmised that another party of survivors must have gone ahead toward the Great Fish River, leaving these men to catch up. Apparently, no one ever returned to this boat, and the Eskimos said there was only one dead man on the ship after it was pushed ashore by the ice. M'Clintock wrote that all men in the group had greatly overestimated their strength and the distance they could travel. He concluded that they had headed back to the ship when they realized the distance was too great for the provisions they brought.

Neither M'Clintock nor Hobson found the least trace of either ship. M'Clintock found a scattering of equipment and supplies that had been discarded along the way as men began to realize they were carrying useless dead weight. He found cooking stoves, pickaxes, shovels, iron hoops, canvas, four feet of copper lightning conductor, pieces of curtain rods, and a medicine chest with vials of well-preserved medicines. A pile of clothing four feet high had been discarded, and M'Clintock's party looked in the pockets of every item of clothing but found nothing. In early June, M'Clintock had to end his search for traces of the lost expedition before thawing ice left him stranded on King William Island. Thawing ice and snow on land also had made travel more difficult. At that time, he began the 230-mile return journey to the *Fox*. A note found in a cairn told him that Hobson was six days ahead of him and gravely ill, and Hobson's party was rushing back in haste to the *Fox* to put Hobson in the doctor's care.

The rains began, and the sea ice became too slushy for travel. After crossing to Boothia Peninsula, M'Clintock had to continue by the much slower overland approach back to his ship. The other sledge parties arrived at about the same time, and everyone indulged in fresh meat, beer, plenty of lemon juice, and pickled whale skin—another known antiscorbutic.

The efforts of M'Clintock's search parties resulted in the addition of 950 miles of coastline to the Arctic map. This completed the map of the North American mainland coastline and of the Arctic islands to about 77° north latitude. They arrived back in England on September 20, 1859, with a great amount of news about the Franklin expedition. M'Clintock gave credit to Franklin for connecting the known parts of the Northwest Passage, although Franklin's stranded ships were a hundred miles from the known connection to the Passage.

Although M'Clintock brought much information about Franklin, many questions still could not be answered. Much more information came later from an American, Charles Hall, who lived among the Eskimos in 1860–1862, learning to live and speak with them. Hall and others found the Eskimos' verbal histories to be reliable, and regarding the Franklin expedition, their verbal histories in combination with relics and bones filled in many details. They told Hall that they had met Franklin and that he was kind to them. They revealed further that the ship with Franklin (*Erebus*) was crushed by ice, and the crew came ashore with some provisions before she sank. The other ship stayed afloat, but was pushed to the shore by ice and abandoned. Hall said that through his research he could account for 79 of the 105 men who had left the ship.

More details were gradually discovered by other expeditions in later years, but questions remain today. In 1878, Lieutenant Schwatka of the U. S. Army made an overland journey, found more bones and relics, and heard accounts from Eskimos regarding the remains of thirty white men on the mainland. These Eskimos had seen many papers near the bodies, but the documents no longer existed when Schwatka arrived. In 1928–1929, a Canadian government surveyor, Major L. T. Burwash, heard from Eskimos that Franklin had been buried on King William Island in a cement vault, and that some of the other men had been buried and covered with something that became hard like rock. Many guns were fired for these burials. Other skeletons were found in 1931 on an island in Simpson Strait, and more skeletons have been found since then. Even with the many known skeletons, about a quarter of the number in the Franklin expedition are still unaccounted for.[6]

The emerging story from all the accounts remains hazy, but certain elements are firm. One hundred five men from the two ships set off for the Great Slave Lake via Back River. As they reached Terror Bay, only 25 miles from their abandoned ships, the crewmen, weak from hunger and scurvy, had already begun to falter under their heavy loads. They apparently camped for several days to recuperate. Later searchers found several skeletons and two boats in addition to the one found by M'Clintock. Physical evidence suggests that Captain Crozier and any men strong enough to move continued south toward the Back River. As the group slowly moved along the south coast of King William Island, fallen men and abandoned equipment marked their route. At some point, the group split and took different directions, probably hoping that at least one would reach help. Crozier's group of about forty men continued toward the Back River. A second group headed east toward Boothia Peninsula, perhaps hoping to reach open water and intercept a rescue ship, or to find the food cache left by John Ross on the west shore of Prince Regent Inlet. Whatever their expectation, they failed and died.

Meanwhile, Crozier's dwindling party crossed Simpson Strait on the ice and reached the Adelaide Peninsula. Sometime, possibly as much as three years later, Eskimos found bodies of thirty men at a small bay on the north coast of Adelaide, now called Starvation Cove. They had traveled nearly 150 miles from the ships, almost reaching the estuary of the Back River. The Eskimos' account revealed that the corpses had been eaten by humans. They said that bones had been cut through by saws, and skulls had been broken open to remove the brains. Also, the Eskimos found many papers, probably records of the expedition. Having no value to the Eskimos, the papers were discarded and lost forever.

A change of escape plans must have occurred at this time. Crozier and four remaining survivors now headed northeast rather than continuing south along Back River. Eskimos found them on Boothia and helped them survive for at least two years, as they moved with and lived like the native people. Crozier and at least one other man eventually left the safety of the Eskimos and headed again toward the Back River. Among their remaining possessions were guns, ammunition, and a folding rubber boat. Apparently they had little trouble reaching the mouth of the Back River due to their strengthened condition and their new skills of living and traveling like Eskimos.

The men must have known that traveling upstream on the raging and dangerous Back River was beyond the capabilities of their boat and their skills. Abandoning such an impossible journey, they went southeasterly toward Baker Lake where they would follow the Theron River to the Chesterfield Inlet on Hudson Bay. This change of direction was probably made on the advice of Eskimos they met on the Back River Estuary. The distance would be shorter, with a greater possibility of finding help. Eskimos reported seeing two white men in the Baker Lake area sometime after 1852, but they apparently vanished, and neither of them was ever seen again.

The second group that split from Crozier had dwindled down to four that Eskimos claimed to have seen as late as 1862 on Melville Peninsula, west of the Foxe Basin portion of Hudson Bay. Nothing more was seen of them, and their fate, like Crozier's, is unknown. Most of this information is based on Eskimo accounts, with little physical evidence. However, these oral histories should be considered credible. Despite certain discrepancies, there are common elements among them that appear to be based on fact and remain believable.

M'Clintock became recognized as the one who had discovered the fate of the Franklin expedition. John Rae's account had been completely disregarded primarily because of his unbelievable news that the men had resorted to cannibalism.

Eventually, M'Clintock was seen as one of many who contributed detail to the original report of John Rae. However, at the time, Lady Jane Franklin recognized M'Clintock as the one who had validated her husband and presented him with a 3-foot-long silver model of the *Fox*. To each of his men she gave a silver watch engraved with a picture of the *Fox*.

Franklin's starving men purposely had avoided contact with Eskimos who could have helped the stranded mariners survive. The crewmen had persisted in living as Europeans who saw no reason to learn about living in the Arctic environment from an uncivilized people. Most British naval explorers of the nineteenth century also refused to learn from their own experiences, the experience of other explorers, or the Eskimos. Author James Morris attributes this hubris to the success of the British empire in the nineteenth century. He writes, "an assumption of superiority was ingrained in most Britons abroad. . . . [T]he ancient social orders of the subject nations were all too often ignored or mocked."[7]

Applying that idea to the nineteenth-century exploration experience in the Arctic shows that the British failed to recognize an environment in which they could not follow their usual practice of simply transplanting their culture. Few of them ever realized that the indigenous culture held the essential keys to survival in the Arctic.

PART VI

Shifting the Focus to the North Pole Fills

in Vacant Spots on the Map, 1875–1920

To unpathed waters, undreamed shores
—WILLIAM SHAKESPEARE

22

George Nares Maps the North Coast of Ellesmere Island

and Relearns Lessons, 1875

FOR YEARS AFTER the Franklin disaster, the Admiralty and the British public had no taste for further Arctic exploration. The Northwest Passage had been found, though not completely sailed, and too many had died trying. A few naval officers still relished the idea of a new expedition, but their focus had changed to the North Pole—the new prize, of course, must be reached by a British expedition.

Two of these officers, Clements Markham and Sherard Osborn, had served in the flotilla of ships that unsuccessfully searched for Franklin in 1850–1851. They now proposed a British expedition designed to reach the North Pole. Markham confidently claimed that the dangers of the Arctic were a thing of the past and scurvy could now be prevented. Gradually, Markham and Osborn convinced a few influential men to go along with their plan. By 1875, the Admiralty had given approval for an expedition to the Pole, and the ships were ready to sail. Although the Admiralty's official instructions stressed the importance of science in this expedition, the sole object of interest was the Pole. Patriotic enthusiasm among the public and the press ran high. The level of interest could well be compared to the rapt attention of the American public for the first manned moon shot in 1969.

Sir George Nares (1831–1915) led the expedition with two ships, *Alert* and *Discovery*, both equipped with sail and steam. Nares was a career naval officer who, at the age of 45, had no avid interest in returning to the Arctic. His past experience with the Belcher fleet in the search for Franklin was the reason for giving him command of this venture, and he saw it, as career men will, simply as a good career move.

The level of preparation for this expedition showed that the navy had learned nothing about Arctic survival, even after dozens of expeditions, some disasters, and subsequent inquiries. John Rae's extensive testimony on survival in the Arctic apparently left no impression on the men preparing for the Nares expedition. One member of the expedition reluctantly followed Rae's advice and took snowshoes aboard the ship. He was met with laughter from his fellow officers, but in the end he thanked Rae. Rae's advice on the use of igloos was also ignored, and heavy canvas tents were loaded on board. Rae had told the inquiry that an igloo big enough for five men could be built in an hour at the end of each day of overland travel. Tents were extremely heavy and difficult to fold when stiff with ice, and they added immense weight for sledging parties. Nevertheless this expedition chose tents. The expedition also took fifty-five dogs, but only one experienced dog-team driver. The British naval attitude toward sledging was that it was a man's job and somehow undignified to use dogs. The sleeping bags did not conform to Rae's recommendation to use fur lined bags, which he had learned from the Eskimos. Kegs of lime juice later became unusable when the juice froze solid. A rum or gin mixture in the juice would have helped that situation—another lesson learned and ignored.

The ships left Portsmouth in late May 1875 with full expectation of setting the British flag at the North Pole. They passed through Kane Basin and Kennedy Channel and became the first ships to traverse the entire route between Greenland and Ellesmere Island into the Lincoln Sea. At that latitude, they still expected to view the imaginary open polar sea that had been hypothesized for several hundred years. Instead, they saw the vast polar ice pack without end. "We had arrived on the shore of the Arctic Ocean and found it exactly the opposite to an Open Polar Sea."[1] According to untested theory, they thought the polar sea should be ice-free because the saline ocean water, with no large land area nearby, should be undiluted by fresh water, and the continuous summer sunlight should completely melt any winter ice. Now Nares' expedition had reached the highest latitude (82°24′ N) that any ship had ever sailed, and Nares concluded correctly that no land existed between their position and the North Pole (Chapter 20, figure 20.1).

Nares immediately saw that the rough surface of the polar ice cap would not allow easy travel for the heavy sledges. "The floes formed as rough a road for sledge traveling as could be imagined."[2] Most of the pressure ridges would require strenuous labor to cross, unloading and reloading the sledges several times each day, slowing progress to a crawl. On August 31, Nares found an inlet safe from the crushing ice floes and prepared to face the Arctic winter. Like other Arctic explorers, Nares described with awe the power of the immense blocks of ice that coalesced and compressed together forming a solid wall across the inlet, protecting them from the crushing movements of ice floes beyond. Preliminary sledge trips went out in

September to set out provision depots for the big push the following spring. Even these short sledge trips proved to be almost beyond endurance. Without snowshoes, the men pulling the sledges sank in the snow to their knees with each step. The men returned with frostbite so severe that some toes or fingers required amputation.

Nares planned to send several sledge parties out in the spring. One would explore the north coast of Greenland, one the north coast of Ellesmere Island, and one, led by first mate Markham, would strike out across the ice for the Pole. Each party would take a boat as a precaution against being stranded by shifting ice floes. Spring came and the push began on April 3, 1876. Their difficulties began almost immediately. The poor choices of clothing, bedding, and shelter and the lack of snowshoes subjected the men to miserable conditions in which it was impossible to remain dry and warm. The heavy loads of wet and frozen tents and boats brought such exhaustion that the food rations were insufficient to restore their energy. As if that were not enough, the men began to come down with scurvy early in the trek. At first, a few men complained of severe pains and fatigue, but they failed to recognized the first symptoms of scurvy. "[O]ne of my sledge crew, told me in answer to my inquiry as to why he was walking lame, that his legs were becoming very stiff; he had spoken to Dr. Coppinger about them, but attributing the stiffness and soreness to several falls that he had had, he did not think much of it, before that officer's departure; now, however, there was pain as well as stiffness, and both were increasing."[3] Within three days, some of the men had to be carried on the sledges, adding to the load for the rest. By April 19, Markham had to abandon one boat to compensate for the extra load of sick men.

On April 24, after three weeks of travel, Markham knew he must return to the ship while there were still enough healthy men to pull the sick back to the *Alert*, 40 miles away. The race for the Pole had become a retreat for survival. Markham sent one man ahead to the ship for help, and in his rush he reached the ship in twenty-three hours so exhausted he could barely stand. Nares sent a rescue party by dog team. By the time Markham's party arrived back to the ship, only three of fifteen men could still walk.

When Nares determined the sad condition of Markham's group, he realized the westward mapping group led by Lieutenant Aldrich must be in the same condition. He immediately sent a relief party. They found only four men in Aldrich's party able to work the sledges. Two stumbled and crawled along behind the sledges, and four more were so weak they had to be carried on the sledges. Nevertheless, Aldrich had traversed over 200 miles of unmapped coastline to Alert Point on the north shore of Ellesmere Island.

Conditions were not much better on board the *Alert*. By the end of June, only nine men out of a party of fifty-three could do the work necessary on the ship. The other sledge parties suffered much the same as Markham's group. Nares made the observation that the weaker men first showed signs and the heartier men later. He also noted, however, that fewer officers suffered from scurvy. Nares attributed this to the fact that some officers had brought private food supplies including hams, vegetables, milk, cheese, and eggs. Knowing this, he donated his own supply of food to the sick men.

The Nares expedition had planned to stay in the Arctic until 1877, but now they realized they had ceased to function as an expedition and must return to England. It was July, and the *Alert* was still icebound. Nares' crew blasted a channel with explosives to reach open water and sail home. They arrived in England on November 2, 1876, and recriminations soon began. The press laid much of the blame on the Admiralty, but Nares unfairly received the brunt of criticism for having the audacity to abandon the race to the Pole. The press called for a court-martial and an inquiry into Nares' mismanagement. Fortunately, there was no legal action against Nares, but the Admiralty had lost its taste for polar expeditions altogether. This was the last British venture into the Arctic until the Watkins party wintered on the Greenland icecap in 1929–1930.

Public opinion had remained high for some previous Arctic expeditions that had accomplished little, but too much had been expected of poor George Nares. In the end, supporters pointed to the scientific data as the main achievement of the expedition. The general public, however, misunderstood the importance of reams of scientific observations and data that expanded the biological, geological, astronomical, and climatological knowledge of the Arctic. The scientist members of the expedition produced at least forty scientific papers from their data. The maps made of the north coast of Ellesmere Island remained a primary source of information until aerial photographs became available in the third decade of the twentieth century. Not least in importance, the Nares expedition ultimately contributed to Canada's claim on all the Arctic Islands.

23

Otto Sverdrup Maps an Immense Area, 1898

———

THE NORWEGIAN EXPLORER, Otto Sverdrup (1854–1930), made a giant leap in a new direction. From 1898 to 1902, he went to the Arctic solely for the purpose of mapping and collecting scientific information. His expedition involved no search for the Northwest Passage, no rush to the North Pole, and no search for an open polar sea. Sverdrup regarded the race to the Pole as something akin to an international sports event and had no desire to participate. There was no glamour to his project—just mapping and science. His ultimate destination was the great area west of Ellesmere Island that had been ignored by all previous expeditions. The area offered no potential route to the Orient and, being ice-locked most of the time, had to be surveyed primarily by sledge. Although Sverdrup never became well known among the public, he added some 100,000 square miles of land area with over 1,700 miles of coastline to the Arctic map. No other single expedition came close to achieving that much newly mapped territory.

Sverdrup was already well known among Arctic travelers as second-in-command on Fridtjof Nansen's famous drift with the ice across the Polar sea in the icebound *Fram* from 1893 to 1896. Nansen had deliberately allowed his ship to become locked in the polar ice cap to observe its movement. When Sverdrup was asked if he would lead an expedition in 1898, he noted, "There are still many white spaces on the map which I was glad for the opportunity to color with the Norwegian colors, and thus the expedition was decided."[1] Canada claimed rights to all the islands already discovered by the British, but other countries still viewed the unclaimed areas as open for whalers and the claims of explorers. Thus Sverdrup acquired land for Norway.

On this new venture, Sverdrup commanded the *Fram* with a crew of fifteen well-chosen men. When word spread that Sverdrup was planning another expedition, applications came in from many parts of the world. The group he selected, besides experienced seamen to operate the ship, included a Norwegian army cartographer, a Swedish botanist, a Danish zoologist, a Norwegian physician, and a Norwegian geologist. This was a model expedition, precisely planned and executed. In June 1898, the expedition departed from Kristiansand, on the south coast of Norway, and headed for the west coast of Greenland.

Sverdrup sailed north along the west coast of Greenland and became locked in ice for the winter of 1898–1899 on the east coast of Ellesmere Island. In the spring they sledged with dogs to the west side of Ellesmere, where he could see the uncharted island he named Axel Heiberg Land for the Norwegian Consul who helped equip the expedition.

Once back on the *Fram*, Sverdrup took the ship south and sailed into Jones Sound, where he anchored in Harbor Fjord on the south shore of Ellesmere Island. There he spent the winter of 1899–1900. The following summer, he moved to Goose Fjord at the west end of Jones Sound and completed a detailed map of the south coast of Ellesmere Island. From this time through 1902, the *Fram* was frozen in place, and the remainder of their mapping and scientific work covering the immense area to the northwest was done by dog sledge as shown in Chapter 20, figure 20.1.

During that time, they mapped all the west coast of Ellesmere Island, the entire island of Axel Heiberg Land, and two islands, Ellef Ringnes and Amund Ringnes, named for two brothers who owned a brewing company and who also had helped equip the expedition. Near the end of the expedition, Sverdrup reached the north-ernmost point on the west coast of Ellesmere, which he named Lands Lokk (Land's End). The north coast of Ellesmere had previously been surveyed by Pelham Aldrich in 1878 while looking for a suitable starting point for the dash to the Pole as part of the Nares expedition. Robert Peary traversed the north coast in 1906 for the same reason.

In 1904, with input from his colleagues, Sverdrup wrote a two-volume narrative of the expedition, with an English translation edition titled *New Land*. The tone of the writing is upbeat and gives the impression of competent and brave men who relished their experience. He told the exciting experience of his second-in-command, Victor Baumann, on a survey of the west coast of Ellesmere Island. Baumann had taken some time to hunt musk oxen and suddenly found himself the target of a charging herd of thirty animals. The nature of the surrounding terrain and the speed of the charge left him no escape. His only options were to stand and be crushed or to counter-charge the herd.

> The animals suddenly became aware of me, and wheeled right round and headed straight for me at a full gallop. So close on each other were their horns that they seemed to form an unbroken line. The animals sunk their heads until they almost touched the ground, and they snorted, blew, and puffed like a steam engine. There was no time for prayer or reflection. If this was to be the end of me then, in Heaven's name, let me rush into it rather than standing still. So, with a horrible yell, and waving my arms, I charged the line. This did some good, for as I came nearer I saw the rank open, and I ran straight through it. The nearest animals were not a yard from me. Before I had time to think the whole herd wheeled round, coming towards me again, and I once more charged the line. As before the ranks opened, and I slipped through unscathed.

Next the herd broke into smaller groups and began to come from all sides. Poor Baumann felt doomed this time. His dogs arrived on the scene in the nick of time and dispersed the herd long enough for him to escape.[2]

After returning to Norway, Sverdrup's team of scientists wrote thirty-five scholarly publications. Maps made by Gunnar Isachsen, Sverdrup's cartographer, became widely known. Sverdrup was given the Grand Cross of the Royal Norwegian Order of St. Olav, and a medal from the Royal Geographical Society. Norway honored Sverdrup for his achievements but took little interest in their rights on the land he had mapped. The Canadian government, however, had a great interest in claiming all the lands of the Arctic Archipelago. In 1930, the Norwegian government ceded their claim of the islands to Canada. Sverdrup sold his maps and notes to the Canadians for $67,000 (Canadian dollars), a handsome sum in 1930. Two weeks later, Sverdrup died at the age of 76.

ROBERT PEARY'S EXTENSION OF ALDRICH'S SURVEY

Although Robert Peary was famous for his efforts to reach the North Pole, he made important contributions to mapping the northernmost coasts of Ellesmere Island. In 1906, Peary, after suffering a setback in his latest effort to reach the Pole, decided to salvage his reputation by traversing westward along the north coast of Ellesmere Island to extend the survey made by Lieutenant Aldrich in 1876. Most of Peary's journey along the north coast retraced the route Aldrich had taken, but he went farther and connected the unmapped 95-mile segment from Alert Point to Lands

Lokk, the most northeasterly point reached by Otto Sverdrup in 1900, now called Kleybolte Peninsula.[3] Peary noted that it had been "particularly gratifying to close the gap in the coast line between Aldrich's and Sverdrup's farthest points."[4]

In 1898, Peary and Sverdrup met in a brief episode that gives good insight to the differing characters of these two explorers. Sverdrup had no objective other than mapping new lands and reporting on their terrain and resources. He did not regard the race to the Pole as a serious endeavor. Peary, on the other hand, had a driving ambition to become famous by being first to reach the North Pole. Whether he actually achieved that goal is still uncertain. In a brief meeting of only a few minutes on the east coast of Ellesmere Island, Peary, on an earlier expedition, happened onto Sverdrup's encampment. Assuming that Sverdrup was a competitor in the race to the Pole, Peary was cool and would not discuss anything lest he give away his intentions. To Sverdrup's surprise, Peary left abruptly and doubled his efforts northward to make certain that Sverdrup could not steal his glory.

24

Vilhjalmur Stefansson Maps New Islands, 1913

AT THE BEGINNING of the twentieth century, the era of mapping by ship and dog sled was drawing to a close. The British had lost interest in the Arctic after the Nares disaster in 1875–1876, and the Canadians and Americans had moved to center stage. The North Pole still attracted men like Robert Peary and Frederick Cook who each wanted to be the first to reach this elusive target. But the task of finishing the map meant discovering and mapping the remaining islands of the North American Arctic. It was uncertain if more islands existed and, if so, where one should look for them. One man, Vilhjalmur Stefansson (1879–1962), stepped forward to finish the map.

Stefansson, born in Canada to Icelandic immigrant parents, was trained as an anthropologist with a focus on native people of the Arctic. Through his study of the Arctic, he became skilled in Arctic survival, exploration, and mapping methods, and these abilities helped him convince the Canadian government to support a major exploration project in 1913. Unfortunately, the project began badly in August when his expedition ship, *Karluk*, commanded by Captain Robert Bartlett (1875–1946) of Newfoundland, became locked in ice after steaming through Barrow Strait and eastward past Point Barrow, Alaska. Stefansson left the ship and walked over the ice 25 miles to Point Barrow.

While he was gone, the *Karluk* became free of ice and continued, but instead of staying near shore and open water, Captain Bartlett steamed north hoping to find an open lead in that direction—ignoring the general knowledge that open leads are more likely to occur near shore. Instead of finding a clear passage, the *Karluk* again became locked in ice—this time permanently. The ill-fated *Karluk* drifted westward with the shifting ice pack for five months from August to January 1914, when it finally was crushed and sank near Wrangle Island off the Siberian coast. Twenty-two men, one woman, two children, and sixteen dogs began life on an ice floe. The survivors salvaged the ship's supply of canned pemmican, hardtack bread, canned milk, and tea. Their dietary preference for bread soon depleted the hardtack, and pemmican, supplemented by hunting, kept them alive. A party of four men tried to make a 300-mile trek to the mainland and were never seen again. The rest of the survivors walked the 80 miles to Wrangle Island in mid-March, traveling twenty-four days over the ice and hacking through a massive ice ridge 75 to 100 feet high. Captain Bartlett with one man made his way across ice 200 miles to the Siberian mainland, then eastward along the coast to the Bering Strait. They went on by ship to Alaska where a rescue ship was arranged. The survivors' long ordeal on Wrangle Island was finally ended when the rescue ship arrived in early September 1914, but eleven of their original party had died. Captain Bartlett was the hero of the day after his six-month-long effort to rescue the survivors. Stefansson, on the other hand, met some criticism for having left the ship as soon as it became icebound. He explained that he had left the ship with five men, two of whom happened to be his exploration party, for a ten-day hunting trip, and the ship had drifted away while he was gone. When he discovered he could not return to the *Karluk*, Stefansson hiked to Point Barrow, Alaska.

After wintering at Point Barrow, Stefansson, with two men, began their long lasting trek of discovery, mapping, and collecting scientific data in March 1914. They reached Banks Island, where they established a base camp for the winter of 1914–1915. In the spring of 1915, they began their exploration for new land by going westward onto the ice of the Beaufort Sea. See Chapter 20, figure 20.1 for a map of Stefansson's multiyear mapping expedition. Stefansson had been told by Eskimos that there would be no seals over the deep water seas. He had also heard this information from whalers and from seasoned Arctic experts such as Robert Peary and Fridtjof Nansen. Stefansson assumed that none of those informants, including the Eskimos who always stayed close to land, had ever actually tried to hunt seals in deep water areas. His hunch was correct, and he discovered that seals existed far from land in sufficient numbers to sustain his party of three men and six dogs indefinitely. He commented that two seals per week were enough for their needs. Stefansson carried no provisions, so they essentially lived off the sea for five years.[1] In his studies of Eskimos, he found that for six to nine months their diet was nearly 100 percent meat and fish, with no carbohydrates. By adapting to this diet, his exploration party proved that people accustomed to a European diet could adapt to a diet of only fat and meat, including the entrails.

Stefansson described in great detail the technique for hunting seals that he had learned from the Eskimos.[2] To approach a seal basking on the ice, one must begin crawling and wiggling in seal fashion from about 300 yards away. By keeping such a low profile, the hunter will probably be unnoticed by the seal until about 200 yards away. The seal is napping but raises its head every few minutes, scanning the area for polar bears. The hunter must crawl forward when the seal's head is down and stop when the seal looks up. When the seal sees the hunter's approach, the hunter must stop, turn his body to the side and make seal-like motions, such as raising his head, or rolling on his back and raising his knees toward his chest briefly to imitate a scratching movement that seals make. If the hunter were to stay perfectly still, the seal would become suspicious and slide into the nearby water. Eventually, the hunter comes within 50 to 100 yards of the seal and can easily make a successful rifle shot to its head. In this way, Stefansson managed to live in the Arctic subsisting on the land and sea.

Finding no new lands to the west, Stefansson went back to Patrick Island to complete the survey done by Leopold M'Clintock in 1853 as part of the Belcher expedition in search of Franklin. Stefansson's party found no game on Patrick Island, but they still found seals on the ice. They completed the mapping and returned to base camp on Banks Island for the winter, where they were met and resupplied by a ship.

The next spring, 1916, the three men began a grand loop that included surveys of Brock Island and Borden Island, which were the first new and uncharted islands they had encountered. Stefansson claimed these islands "in the name of King George V on behalf of the Dominion of Canada." The process of making a formal statement and a map validated the claim. Later in 1947, aerial photographs showed that the land mapped as Borden Island was actually two islands; the second one became Mackenzie Island, named for the then prime minister. The 1916 survey continued northeast along the ice looking for more islands. They found two more unmapped islands, Meighen Island and Lougheed Island, before returning to their base camp. On the return to Banks Island, they passed by Mercy Bay where McClure and crew on the *Investigator* had been icebound for two winters sixty-three years earlier. They found some articles from the McClure expedition and updated the map of the area.

In the spring of 1917, Stefansson's party pushed northward onto the ice of the Arctic Ocean, but finding no additional islands, returned to base camp, ending the field work of the project. Stefansson became well known among Arctic professionals by demonstrating that expeditions could be sustained for extended periods of time on the ice, even though some of the small islands lacked game. He had found several new islands and compiled a great amount of scientific data on ocean depths, ice movements, winds, and currents. Stefansson gave high praise to his two companions, Storker T. Storkerson and Ole Andreason, who spent so much time with him. "They are as well suited for this work as it is easy to imagine. Neither of them worries or whines and both are optimistic about the prospects. This last is important. Traveling with an empty sled and off the country is no work for a pessimist."[3]

After Stefansson's discoveries, the map of the Canadian Archipelago was essentially complete, and extensive mapping expeditions using dog sledges came to an end. The era of ships intentionally becoming icebound for the purpose of exploration and mapping expeditions gave way to the excitement and practicalities of aviation. A party could be airlifted to a site in a day rather than months, and the plane could return to resupply them at regular intervals. Radio contact now kept workers in the Arctic in constant touch with the outside, so help could arrive on short notice when needed. In short, everything changed soon after World War I.

Just six years after Stefansson's expedition, the first aerial photographs in the Arctic were made north of Norway over the island of Spitzbergen in 1923. In the next year, a complete mapping survey of Spitzbergen was done by aerial photography.[4] Canada made its first aerial photographs over Ottawa in 1920. Aerial mapping continued across Canada and the archipelago, remapping and refining work begun when Samuel Hearne and Alexander Mackenzie discovered the north edge of the continental mainland in the eighteenth century. Mapping by aerial photography finally reached the remotest Arctic islands in mid-twentieth century. Lougheed Island and Borden Island were first covered by aerial photography in 1959. The next year, Patrick Island received its first aerial coverage.[5]

The twentieth century saw great strides in the ability to map not only coasts, but all aspects of the environment, including atmosphere, land, and sea. In the 1970s, excellent images taken from the LANDSAT satellite made the mapping of large areas possible, providing detailed thematic maps of geology, vegetation, and ice. Later RADARSAT made it possible to continue mapping through long months of Arctic darkness and cloud cover, to differentiate types of ice, and to track drift ice all year.

25

A Few Final Thoughts

BRONOWSKI AND MAZLISH wrote that the center of the intellectual universe shifted from the northern Italian states to France, England, and Holland in the late sixteenth century. The tenor of the times encouraged extensions of experience and investigation of the imaginary to solve real problems. With expanded thinking, they could imagine new lands to claim and passages through continents as a means to fame and prosperity. Bronowski and Mazlish explained that the various reasons for this change were closely connected to the breakout from a focus on the Mediterranean into the Atlantic and toward North America.[1] As Spain and Portugal declined during this period, England rose in prominence under Elizabeth I and became adept at claiming new areas. Europeans regarded their new discoveries as possessions throughout the four hundred years spent exploring and mapping the boundaries of North America. They had maps and descriptions showing they had performed formal rites of possession, and they often established settlements to cement the fact. As described in Chapter 1, the key elements were a map, a narrative, and a memento from the new place. Frobisher's rocks and kidnappings were his mementos.

We have seen that many expeditions suffered great hardship and even death for the sake of completing these maps, but nowhere were these hardships more severe than in the Arctic. Prior to exploring the far north, most voyages to North America were completed within one year. For longer expeditions, there were options for restocking food supplies by hunting and fishing. However in the Arctic, game was often scarce and the short sailing season allowed little time before the ice closed in.

Only after ships became larger and stronger could they pack enough provisions and fuel to remain and allow themselves to become icebound, resuming work in the next year. All went well if the ship could free itself from the ice in the next summer, but sometimes they were forced to stay another year or more, depleting their provisions. When this happened, the possibility of death by starvation or scurvy became serious. Much has been written on the reasons for European explorers dying while the native Eskimo population lived well in that harsh environment.

We can see a first stage of Arctic exploration during which explorers were so unprepared for the harsh environment that they made only tentative probes, intending to return to a milder climate by the end of September. Frobisher and Davis are typical of this period. Henry Hudson wintered over, but with dire consequences. In the second stage, ships went to the Arctic with the intention of wintering at least one year, often more. Winter in this stage was something to be endured until work could begin again in the spring or summer. Ships became frozen into inlets and prepared for the icy blasts. Edward Parry was one of the first to test this approach, and in doing so made a record penetration into the Canadian Archipelago.

In the 1840s, explorers began a third stage when they recognized the need for using the dog sledges to traverse areas their ships could not reach. When this happened, the men saw winter in a different light. The sledges could travel more easily over a frozen sea than on land, returning to the ship when thawing began. With this approach to surveying, sledge parties could leave the frozen ships in February when daylight returned and work until signs of thaw began to appear. M'Clintock and Rae were early practitioners of winter sledging. Another aspect of this stage was the increased dependence on survival by hunting seals, caribou, and muskoxen. Rae and Stefansson relied totally on game, carrying no food in the sledge except what they hunted along the way. Stefansson's companion, Storkerson, wrote, "Repeatedly we were down to our last meal, but always before it was gone we had a chance to hunt and so replenish."[2]

Vilhjalmur Stefansson learned from the Eskimos and the experience of Hudson Bay Company employee, John Rae, who had also learned survival methods from the Eskimos, and Rae made survival recommendations to the Admiralty. Unfortunately, his suggestions were mostly ignored. The fundamentals were simple: keep clothing dry, learn to build snow houses, learn to hunt seals, wear clothing made of animal skins as the Eskimos did, and split large groups into smaller groups and spread out over the land to better utilize the available resources. John Rae and his party had managed to live off the land in the Arctic for extended periods of time. He had eighteen men in the vicinity of Repulse Bay, which was believed to have less game in fall and winter. They took only a small amount of food with them. Rae's report shows that by October they had killed 130 caribou, 200 partridges, and many salmon. He built a snow house for storage that was well stocked with food.

Yet Rae was held in low esteem by the English for living like a menial, i.e., doing his own work, and for living like a savage in skins and snow houses. Stefansson saw this attitude as the culture of the British gentry when he wrote, "Just as with fox hunting for the British gentleman, in which the prime objective is not killing the fox, but the proper observance of form during the pursuit and kill, also, explorations should be done properly and not evade the hazards of the wilderness by the vulgarity of going native. They must face the dangers of the Arctic or other wilderness in the proper manner." He wrote further "that the crews of the *Erebus* and *Terror* perished as victims of the manners, customs, social outlook, and medical views of their time."[3]

It should be noted that the attitude of exploring the Arctic as "gentlemen" existed primarily in the Admiralty and among officers of the Royal Navy. Hudson Bay Company employees, who were also primarily British, survived well on Arctic expeditions by depending heavily on fundamentals described by John Rae. Norwegian, Canadian, American explorers, and the voyageurs all coped with the Arctic by following the Eskimo ways. Some Hudson Bay Company employees ridiculed the English naval explorers because of their unwillingness to give up comforts while in the wilderness. The widely known Chief Factor of the HBC, George Simpson, is quoted as saying, "Lieutenant Franklin, the officer who commands the party [referring to Franklin's first expedition], must have three meals per day, tea is indispensable, and with the utmost exertion he cannot walk above eight miles in one day. It does not follow if these gentlemen are unsuccessful that the difficulties are insurmountable."[4] Despite these criticisms, the British naval officers' devotion to duty and acts of heroism were a source of pride to their nation.

After 1860, much of the Royal Navy had some experience in the Arctic, but the lessons of survival had not yet overcome tradition. The tragedy of the Nares expedition in 1875 was that they professed to have learned all the necessary skills to conquer the hazards of the Arctic, but they actually regressed thirty years, with starvation and scurvy decimating their party. It is astonishing that despite their hardships and suffering, Arctic explorers always managed to stay focused on their objective—collecting information for maps, narratives, and science. In each of these categories they produced valuable results.

In the fifty years from 1492 to 1542, the map of the east coast of North America was extended from Florida to Labrador, and then the pace slowed considerably. Two hundred seventy-six years later, in 1818, the map of the North American Arctic consisted of a point on the east coast, a point on the northwest coast of Alaska, and two points in between—at the mouths of the Mackenzie and Coppermine Rivers. In another forty years, after many expeditions, nearly all the Arctic islands and waterways were mapped, much of it done by men on foot with dog sledges. M'Clintock estimated that the foot travel by all sledging parties alone amounted to 40,000 miles. It is important to remember that these expeditions all collected scientific information during their long treks on the ice. Coolie Verner wrote that the real achievement was more than adding many miles of coastline to the maps; it was the simultaneous acquisition of scientific information in the areas of natural history, geology, climate, winds, currents, Eskimo life, and survival skills for long periods in the intense Arctic.[5]

Trevor Levere points out the immediate value of the reams of scientific data the explorers collected. The need for information to aid in navigation made scientific observation essential. Important information included maps, landforms, ice conditions, currents, magnetic field behavior, and food sources. Little scientific data collection took place on expeditions before the middle of the eighteenth century but became paramount in the nineteenth and twentieth centuries.

Scientific data collected by explorers also had value beyond the immediate needs of navigation. Much of it was used to advance understanding the environment. During the one hundred years after 1818, there were about two hundred expeditions to the Canadian Arctic; almost all of them expanded the knowledge and understanding of the Arctic, and 40 percent of them produced results that led to scientific publications. Most of the commanding officers of the Royal Navy during that period were admitted to the Royal Society of London for their contributions to science.[6] The Royal Society worked closely with the Admiralty providing advice on the collection of scientific data during exploration expeditions. Naval officers were trained to observe and record scientific information, the proper use of instruments, and collection of specimens. Trained civilian scientists were not asked to join naval exploratory expeditions, so typically the ship's surgeon became the primary science observer on board.

The nineteenth century thus became a milestone not only in mapping the Arctic, but also in expanding scientific knowledge of flora, fauna, geomagnetism, geology, hydrography, and ethnology. The scientific study of the Arctic is becoming even more important in the twenty-first century as ice caps begin to recede and the potential grows for ice-free travel through the Canadian Archipelago. The scientific data collected by nineteenth-century explorers stands as a valuable benchmark against which today's conditions can be compared. Recent data provides indisputable evidence of a warming trend that is causing a rapid and shocking decrease in the extent and duration of Arctic ice. This situation would have benefited early explorers like Franklin, but as human activities such as sea transport and habitation continue to expand into the new ice-free areas, there will be profound and long-lasting effects on wildlife and the fragile environment of the Arctic.

GLOSSARY

ARMILLARY SPHERE A model of the celestial globe constructed from rings representing the equator, the tropics, and other celestial circles, and able to rotate on its axis.

BARK (BARQUE) A sailing ship, typically with three masts, in which the foremast and mainmast are square-rigged and the mizzenmast is rigged for a lateen sail.

CAPTAIN OF THE SHIP The person in command. He may or may not have been the ship's master or have the naval rank of captain.

CARAVEL A small ship of the fourteenth and fifteenth centuries averaging 70 to 80 feet long with two square rigged masts and a lateen rigged mizzen mast. It was valued for its speed, maneuverability, and capability for sailing close to a headwind. Columbus's *Nina* and Cabot's *Mathew* were caravels.

CARRACK Similar to the caravel, but larger and having high castles both fore and aft. Sails were rigged similar to the caravel.

CASTLE The high structure built on one or both ends of a ship.

FURL To roll up and secure the sail to the yard.

GALLEON A ship developed from the carrack and used extensively for trading, exploration, and warfare. It was valued for its handling and maneuverability in all weather.

HALYARD A rope used for raising and lowering a sail, spar, flag, or yard on a sailing ship.

ICE FLOE A sheet of floating ice composed of frozen seawater. It is distinguished from an iceberg, which is freshwater ice calved from glaciers.

JIB A triangular sail set on a line, rather than a yard, ahead of the foremast.

JUDDER Refers to the shaking and vibration of a rudder in heavy seas. The vibration may cause the steering wheel to shudder in the helmsman's hands.

LATEEN SAIL A triangular sail attached to a long yard set obliquely to the mast with the forward end down.

LEECHES The vertical edges of a square sail.

MASTER OF THE SHIP An officer acting as navigator and ship handler who maneuvered the ship as directed by the captain. In later usage, the captain was also the master.

MERIDIAN An arc of constant longitude passing through a given location of the surface of the Earth from the North to the South Poles.

MIZZEN MAST The mast aft of the main mast.

PARALLEL A circle of constant latitude passing through a given location at a right angle to a meridian.

PINNACE A small boat, with sails or oars, often used as a tender to a larger vessel.

REEF THE SAIL Refers to shortening a sail by an amount appropriate to the strength of the wind.

RHUMB LINE A line drawn on a globe or map with constant direction. A course set by compass forms a rhumb line.

SHALLOP A small undecked ship under twenty-five tons. It may be rowed or rigged for sail. A shallop might be towed behind a larger ship for emergencies.

SHROUDS Ropes rigged to give masts lateral support.

SPAR A general term referring to yards or booms.

WAIST The central part of the upper deck of a ship.

YARD A large cross piece attached to the mast. Sails are hung from the yards.

ZENITH The point in the sky directly overhead to an observer on earth.

A CHRONOLOGY OF SELECTED EXPEDITIONS
TO NORTH AMERICA

∽ ———————————————————————————————

Year	Explorer	Ship(s) Sailed	Place(s) Reached
1497	John Cabot	Matthew	Newfoundland
1500	Corte Real, 2nd voyage	2 ships	Greenland, Newfoundland
1524	Giovanni Verrazzano	Dauphine	North Carolina to Newfoundland
1534, 1535, 1541	Jacques Cartier	Grande Hermine, Emerillon, Petite Hermine	St. Lawrence River to site of Montreal
1576, 1577, 1578	Martin Frobisher	Gabriel	Frobisher Bay of Baffin Island
1585, 1586, 1587	John Davis	Sunneshine, Mooneshine	Cumberland Sound of Baffin Island
1609	Henry Hudson, 3rd voyage	Half Moon	Hudson River
1610	Henry Hudson, 4th voyage	Discovery	Hudson Bay

(continued)

Year	Explorer	Ship(s) Sailed	Place(s) Reached
1612	Thomas Button	Resolution	Hudson Bay west coast
1615	Robert Bylot, William Baffin	Discovery	Baffin Island, Hudson Bay
1631	Luke Foxe	Charles	Foxe Basin of Hudson Bay
1631	Thomas James	Henrietta Maria	Western Hudson Bay and James Bay
1728	Vitus Bering	St. Gabriel	Through Bering Strait
1741	Vitus Bering, Aleksei Chirikov	St. Peter, St. Paul	Southeast coast of Alaska
1771	Samuel Hearne	Overland journey	Followed Coppermine River to ocean
1778	James Cook	Resolution	Icy Cape, north coast of Alaska
1789	Alexander Mackenzie	Overland journey	Followed Mackenzie River to ocean
1818	John Ross, 1st expedition	Isabella, Alexander	Lancaster Sound
1819–1922	John Franklin, 1st expedition	Overland journey	Turnagain Point
1819–1920	William E. Parry, 1st expedition	Hecla, Griper	Winter Harbor, Melville Island
1821–1823	William E. Parry, 2nd expedition	Fury, Hecla	Fury and Hecla Strait
1824–1825	William E. Parry, 3rd expedition	Fury, Hecla	Prince Regent Inlet
1825–1828	Frederick Beechey	Blossom	Supply ship to Bering Strait
1825–1827	John Franklin, 2nd expedition	Overland journey	North coast via Mackenzie River
1829–1833	John Ross, 2nd expedition	Victory	Prince Regent Inlet
1833–1835	George Back	Overland journey	Great Fish River
1837–1839	Peter Dease and Thomas Simpson	Overland journey	Extended north coast maps
1845–1848	John Franklin, 3rd expedition	Erebus, Terror	King William Island

Year	Explorer	Ship(s) Sailed	Place(s) Reached
1846–1847	John Rae	Overland journey	Fury and Hecla Strait
		START FRANKLIN SEARCH	
1847–1849	John Rae, John Richardson	Overland journey	Mackenzie River to Coppermine River
1848–1849	James Ross	Enterprise, Investigator	Prince Regent Inlet
1850–1851	Horatio Austin	Resolute	Barrow Strait and sledge journeys
1850–1851	Erasmus Ommanney	Assistance	Barrow Strait and sledge journeys
1850–1851	John Cator	Intrepid	Barrow Strait and sledge journeys
1850–1851	Sherard Osborn	Pioneer	Barrow Strait and sledge journeys
1850–1851	Edwin De Haven, 1st Grinnell	Advance, Rescue	Joined Austin at Beechey Island
1850–1851	John Ross	Felix, Mary	Joined Austin at Beechey Island
1850–1851	William Penny	Lady Franklin, Sophia	Joined Austin at Beechey Island
1850–1851	John Rae	Overland journey	South coast Victoria Island
1850–1854	Robert McClure	Investigator	Mercy Bay of Banks Island
1850–1855	Richard Collinson	Enterprise	South coast of Victoria Island
1852–1854	Edward Belcher	Assistance	Wellington Channel and sledge journeys
1852–1854	Sherard Osborn	Pioneer	Wellington Channel and sledge journeys
1852–1854	Henry Kellet	Resolute	Melville Island and sledge journeys
1852–1854	Francis M'Clintock	Intrepid	Melville Island and sledge journeys
1852–1854	William Pullen	North Star	Supply base at Beechey Island
1853–1854	John Rae	Overland journey	Boothia Peninsula

(continued)

Year	Explorer	Ship(s) Sailed	Place(s) Reached
1853–1855	Elisha K. Kane, 2nd Grinnell	Advance	West coast of Greenland
1857–1859	Francis M'Clintock	Fox	King William Island
	END FRANKLIN SEARCH		
1875–1876	George Nares	Alert, Discovery	North coasts of Ellesmere Island & Greenland
1898–1902	Otto Sverdrup	Fram	Islands west of Ellesmere Island
1906	Robert Peary	Sledges on sea ice	North coast of Ellesmere Island
1913–1918	Vilhjalmur Stefansson	Sledges on sea ice	Brock Island to Meighen Island

Source: Alan Cooke and Clive Holland, *The Exploration of Northern Canada: 500 to 1920: A Chronology* (Toronto: The Arctic History Press, 1978)

NOTES

CHAPTER 1

1. Richard B. Alley. *The Two-Mile Time Machine: Ice Cores, Abrupt Climate Change, and Our Future* (Princeton: Princeton University Press, 2000).

2. George Best. *The Three Voyages of Martin Frobisher: In Search of a Passage to Cathaia and India by the North-West, A.D. 1576–8. With introduction by Sir Richard Collinson* (1867; repr. from 1st ed. of Hakluyt's Voyages, New York: Burt Franklin, 1963), 38.

3. D. Graham Burnett. *Masters of All They Surveyed: Exploration, Geography, and a British El Dorado* (Chicago: University of Chicago Press, 2000), 110–29.

4. Iain C. Taylor, "Official Geography and the Creation of Canada," *Cartographica* 31 (1994): 1–15.

5. D. Graham Burnett, *Masters of All They Surveyed*, 3

6. J. B. Harley. "Deconstructing the Map," *Cartographica* 26, (1989): 13.

CHAPTER 2

1. Peter Gay, *The Enlightenment* (New York: W. W. Norton, 1966), 279.

2. The actual site of Cabot's landfall is in dispute and is claimed by various historians to have been in different locations ranging from Labrador to Nova Scotia's Cape Breton, or even in Maine. The account given here is based on convincing arguments for Newfoundland by Samuel E. Morison. *The European Discovery of America: The Northern Voyages* (New York: Oxford University Press, 1971), 199. Arguments for landfall in Newfoundland are presented again in

Samuel E. Morison, *The Great Explorers: The European Discovery of America* (New York: Oxford University Press, 1978), 58.

3. B. D. Fardy, *John Cabot: The Discovery of Newfoundland* (St. John's, Newfoundland: Creative Publishers, 1994), 46.

4. In 1997, a replica of the *Mathew* sailed from Bristol to Newfoundland to commemorate the 500th anniversary of Cabot's voyage. The carefully researched replica is 78 feet (23.7m) long with a beam of 20.5 feet (6.3m) with a draft of 7 feet (2.1m) and 2,360 square feet (219 m²) of sail—about the same as Columbus's ship *Niña*. Its home today is Bristol harbor.

5. Samuel E. Morison, *The Great Explorers: The European Discovery of America* (New York: Oxford University Press, 1978), 58.

6. Morison, *The Great Explorers*, 59.

7. Cabot's assessment of the suitability of Newfoundland for producing silk and logwood was an optimistic stretch—the logwood tree (*Haematoxylum campechianum*) is native to Central America. Red dye existed in Europe, but it was used only for royalty and ecclesiastical garments. The heartwood of the logwood tree provided a valuable new source of red dye, and shortly after the voyages of Columbus, shiploads of logwood logs worth a fortune began to arrive in Europe. Knowing the value of these logs, Cabot no doubt hoped to create a demand for another voyage.

8. David B. Quinn, "John Cabot," in *Oxford Dictionary of National Biography* (New York: Oxford University Press, 2004).

9. Samuel E. Morison, *The European Discovery of America*, 220.

CHAPTER 3

1. Verrazzano's letter to the king was first published in J. W. Jones, *Divers Voyages Touching the Discovery of America and the Islands Adjacent, in 1582.* First series, no. 7 (London: The Hakluyt Society, 1850). A newer translation is available in Lawrence C. Wroth, *The Voyages of Giovanni Verrazzano, 1524–1528* (New Haven: Yale University Press, 1970).

2. Samuel E. Morison, *The European Discovery of America: the Northern Voyages* (New York: Oxford University Press, 1971), 192.

3. Wroth, *The Voyages of Giovanni Verrazzano*, 136.

4. Mooning, exposing one's posterior, is apparently an ancient and cross-cultural practice. Also, note Lawrence C. Wroth, *The Voyages of Giovanni Verrazzano*, 141.

5. W. H. Goetzmann and G. Williams, *The Atlas of North American Exploration* (New York: Prentice Hall, 1992), 38.

6. Goetzmann and Williams, *The Atlas of North American Exploration*, 40.

CHAPTER 4

1. Meg Green, *Jacques Cartier: Navigating the St. Lawrence River* (New York: Rosen Publishing Group, 2004), 24.

2. Stephen Leacock, *The Mariner of St. Malo: A Chronicle of the Voyages of Jacques Cartier* (Toronto: Glasgow, Brook & Co., 1920), 9.

3. Samuel E. Morison, *The European Discovery of America: the Northern Voyages* (New York: Oxford University Press, 1971), 345.

4. Stephen Leacock, *The Mariner of St Malo: A Chronicle of the Voyages of Jacques Cartier.* (Toronto: Glasgow, Brook & Co., 1920), 78.

5. Leacock, *The Mariner of St Malo*, 93.

CHAPTER 5

1. Peter Earle, *Sailors: English Merchant Seaman 1650–1775*. (London: Methuen, 1998), 22.

2. Twenty pounds in the year 1600 would be the equivalent of approximately $2,400 today.

3. Samuel E. Morison, *The European Discovery of America: the Northern Voyages* (New York: Oxford University Press, 1971), 130.

4. Samuel E. Morison, *The European Discovery of America*, 143.

5. Michael Bartholemew, "James Lind and Scurvy: A Revaluation," *Journal for Maritime Research, National Maritime Museum* (January 2002).

6. Eric Newby, *The Last Grain Race* (Hawthorn, Australia: Lonely Planet Publications, 1999), 95.

7. The size of ships in tuns originally indicated the amount of cargo load they could carry using the wine tun as a unit. The tun was a cask with a capacity of 252 gallons weighing about 2,016 pounds (approximately one ton) when filled. The term ton as a unit of displacement later replaced the tun, which referred to load capacity. Cargos were often carried in 63-gallon hogsheads (one-quarter of a tun), which are easier to load and move around.

8. Even today, Polaris is not precisely on the point of rotation. Corrections of about two-thirds of one degree may be needed depending on time and date.

CHAPTER 6

1. George Best, *The Three Voyages of Martin Frobisher: In Search of a Passage to Cathaia and India by the North-West, A.D. 1576–8, With introduction by Sir Richard Collinson.* (1867; repr. from 1st ed. of *Hakluyt's Voyages*, New York: Burt Franklin, 1963).

2. Sebastian Cabot wrote that Basque and Portuguese fishermen had been cod fishing off Newfoundland years before John Cabot first came there, and that Portuguese fisherman referred to the entire area as *Baccalos,* "Codfish Land." The name exists today as Baccalieu Island off the east coast of Newfoundland. Then, as now, fishermen did not make maps or claim land.

3. Best, *The Three Voyages of Martin Frobisher*, 79.

4. Best, *The Three Voyages of Martin Frobisher*, 25.

5. Best, *The Three Voyages of Martin Frobisher*, 83.

6. Some researchers believe that these Inuits were somewhat familiar with Europeans and their ships. A possible explanation is that earlier contacts had been made with Portuguese ships.

7. George Best, *The Three Voyages of Martin Frobisher*, 92.

8. The Eskimo woman taken back to England became sick and died and was buried in Bristol. The child died later in London and was buried in the same churchyard as the man brought over on the first trip.

9. The fate of the five missing sailors remained a mystery for almost three hundred years. An American, Charles Hall, lived among the Eskimos in the Frobisher Bay area for two years around 1862, learning their language and hearing their oral history. From them he heard that the Eskimos kept five Englishmen prisoner until the English ships left for the last time. The men were released, built a boat from materials left behind, and put to sea. Winter was just beginning, and the men likely died before they could reach a safe haven. Hall, whose reason for visiting the Arctic was to look for evidence of the fate of the Franklin expedition, was the first to record the fact the Frobisher's strait was actually a long narrow bay.

10. There are two reasons ice found at sea would taste fresh. When sea water freezes, at about 28°F, salt is excluded and gradually migrates out of the ice in droplets of brine. Therefore, water melted from older sea ice is drinkable. Also, water derived from icebergs is fresh because the icebergs are derived from calving glaciers, formed from snow, and are fresh from the start.

11. George Best, *The Three Voyages of Martin Frobisher*, 236.

12. Countess of Warwick Island today has the Inuit name Kodlunarn, meaning white men's island, and is a designated historic site. Remains of the stone house and the miner's excavations are still visible.

CHAPTER 7

1. John Davis et al. *The Voyages and Works of John Davis, the Navigator* (London: Hakluyt Society, no. 59, 1880. Reprinted New York: Burt Franklin, 1970, 239.

2. Davis, *The Voyages and Works of John Davis*, 4.

3. Samuel E. Morison, *The European Discovery of America* (New York: Oxford University Press, 1971), 391.

CHAPTER 8

1. Philip Vail, *The Magnificent Adventures of Henry Hudson* (New York: Dodd, Mead and Company, 1965), 4.

2. George M. Asher, *Henry Hudson the Navigator: The Original Documents in which his Career Is Recorded, Collected, Partly Translated, and Annotated*, (New York: Burt Franklin, printed for the Hakluyt Society, 1860), 4.

3. Overfall is a term for turbulence sufficient to cause breaking waves in open water. It may be caused by wind or the meeting two strong currents. e.g. a current exiting Hudson Bay mixing with the south-flowing Labrador Current.

4. Donald S. Johnson, *Charting the Sea of Darkness: The Four Voyages of Henry Hudson* (Camden, Maine: International Marine, 1993), 194.

5. Johnson, *Charting the Sea of Darkness*, 195.

6. W. P. Cumming, R. A. Skelton, and D. B. Quinn, *The Discovery of North America*, (New York: American Heritage Press, 1972), 245.

CHAPTER 9

1. W. H. Goetzmann and G. Williams, *The Atlas of North American Exploration* (New York: Prentice Hall, 1992), 122.

2. George M. Thomson, *The North-West Passage* (London: Secker & Warburg Ltd., 1975) 142.

3. Thomson, *The North-West Passage*, 144.

CHAPTER 10

1. Andrew C. F. David, *James Cook*, Oxford Dictionary of National Biography.

2. John Noble Wilford, *The Mapmakers*, (New York: Alfred A Knopf, 1981), 143.

3. Triangulation surveys begin by carefully measuring a baseline. Each end of the baseline angles are measured to a prominent landmark. By knowing one side and two angles, the length of

the other two sides of a triangle can be computed and thus plot the landmark on the map. Starting from this one triangle, a network of triangles is extended over the area to be mapped, forming a basis for accurate surveys of the land.

4. Since the fifteenth century, some mapmakers had shown a hypothetical Southern Hemisphere continent, *Terra Australis Incognita*, to compensate for the greater area of land mass in the Northern Hemisphere. Such a place had been suggested earlier by Aristotle. This was a matter of symmetry and balance in their minds.

5. For information on the clock and longitude, see Rupert T. Gound, *The Marine Chronometer, Its History and Development* (London: J. D. Potter, 1923), or Jonathan Betts, *Time Restored: The Harrison Timekeepers and R. T. Gould, the Man Who Knew (Almost) Everything.* (Oxford: Oxford University Press, 2006), or Dava Sobel, *Longitude: The True Story of a Lone Genius Who Solved the Greatest Scientific Problem of His Time.* (New York: Walker, 1995).

CHAPTER 11

1. Farley Mowat, *Ordeal by Ice* (Toronto: McClelland & Stewart Ltd., 1960), 185.

2. John Franklin, *Narrative to the Shores of the Polar Sea, 1819, 1820, 1821, and 1822* (London: John Murray, 1828), 1:45.

3. Franklin, *Narrative to the Shores of the Polar Sea*, 2:192.

4. Franklin, *Narrative to the Shores of the Polar Sea*, 2:194.

5. Franklin, *Narrative to the Shores of the Polar Sea*, 2:232.

6. Franklin, *Narrative to the Shores of the Polar Sea*, 2:233.

7. Franklin, *Narrative to the Shores of the Polar Sea*, 2:362.

CHAPTER 12

1. Edward Parry, *Memoirs of Rear Admiral Sir W. Edward Parry* (New York: Bible House, 1858), 127.

2. Parry, *Memoirs of Rear Admiral Sir W. Edward Parry*, 134.

3. William Edward Parry, *Journal of the Third Voyage for the Discovery of a North-West Passage.* (1826; repr., London: Feather Trail Press, 2010), 10.

4. Parry, *Journal of the Third Voyage*, 18.

5. Parry, *Journal of the Third Voyage*, 55.

CHAPTER 13

1. John Franklin, *Narrative of a Second Expedition to the Shores of the Polar Sea in the Years 1825, 1826, and 1827* (London: John Murray, 1828).

2. Franklin, *Narrative of a Second Expedition*, 89.

3. Franklin, *Narrative of a Second Expedition*, 166.

CHAPTER 14

1. John Ross, *Narrative of a Second Voyage in Search of a Northwest Passage, and of a Residence in the Arctic Regions during the years 1829, 1830, 1831, 1832, 1833* (London: A. W. Webster, 1835), 152.

2. Ross, *Narrative of a Second Voyage*, 202.

3. Glyn Williams, *Voyages of Delusion: The North-West Passage in the Age of Reason*, (London: Harper Collins Publishers, 2002). 251.

4. Williams, *Voyages of Delusion*, 253.

CHAPTER 15

1. James Morris, *Heaven's Command: An Imperial Progress* (New York: Harcourt, Brace 1973), 114.

2. Thomas Simpson, *Narrative of the Discoveries on the North Coast of America* (London: Bentley, 1843), 8.

3. Ann Savours, *The Search for the North West Passage* (New York: St. Martin's Press, 1999), 155.

4. Vilhjalmur Stefansson, *Unsolved Mysteries of the Arctic* (New York: Collier Books, 1962), 130.

CHAPTER 16

1. Ann Savours, *The Search for the North West Passage* (New York: St. Martin's Press, 1999), 169.

2. Ann Savours, *The Search for the North West Passage*, 183.

3. For an excellent account of Lady Jane Franklin's role in her husband's career, and especially during the search for his lost expedition, see Ken McGoogan, *Lady Franklin's Revenge* (Toronto: Harper Collins Publishers, Ltd., 2006).

CHAPTER 17

1. Fergus Fleming, *Barrow's Boys* (New York: Atlantic Monthly Press, 1998), 369.

2. Fleming, *Barrow's Boys*, 386.

CHAPTER 18

1. Ann Savours, *The Search for the North West Passage* (New York: St. Martin's Press, 1999), 186.

2. Ann Savours, *The Search for the North West* Passage, 189.

3. Pierre Berton, *The Arctic Grail: The Quest for the Northwest Passage and the North Pole, 1818–1909* (New York: Viking Penguin, 1988), 265.

4. Ken McGoogan, *Fatal Passage: The True Story of John Rae, the Arctic Hero Time Forgot* (New York: Carroll and Graf Publishers, 2002), 190.

5. Berton, *The Arctic Grail*, 267.

6. McGoogan, *Fatal Passage*.

CHAPTER 19

1. Ann Savours, *The Search for the North West Passage* (New York: St. Martin's Press, 1999), 218.

2. Savours, *The Search for the North West* Passage, 222.

3. Schwass-Bueckert, Kate. "Team Finds Lost 1800's Ship in Arctic," QMI Agency, appeared in *Toronto Sun* July 28, 2010.

4. Savours, *The Search for the North West* Passage, 230.

5. See Collinson's introduction to George Best, *The Three Voyages of Martin Frobisher*, vii–xxvi.

6. K. Wood and J. E. Overland, "Accounts From 19th Century Canadian Arctic Explorers' Logs Reflect Present Climate Conditions," *Eos* 84 (2003): 410.

7. T. Agnew, A. Lambe, and D. Long. "Estimating Sea Ice Area flux Across the Canadian Arctic Archipelago Using Enhanced AMSR-E." *Journal of Geophysical Research* 113 (2008) (C10011).

CHAPTER 20

1. Pierre Berton *The Arctic Grail: The Quest for the Northwest Passage and the North Pole, 1818–1909* (New York: Viking Penguin, 1988), 250.

2. Edmund B. Bolles, *The Ice Finders: How a Poet, a Professor, and a Politician Discovered the Ice Age* (Washington, DC: Counterpoint, 1999), 97.

3. Elisha K. Kane, *Arctic Explorations: The Second Grinnell Expedition in Search of Sir John Franklin* (Philadelphia: Childs and Peterson, 1856), 148.

4. Pierre Berton, *The Arctic Grail*, 291.

CHAPTER 21

1. Francis L. M'Clintock, *The Voyage of the* Fox *in the Arctic Seas: A Narrative of the Discovery of the Fate of Sir John Franklin and His Companions* (Boston: Ticknor and Fields, 1860), 119.

2. Vilhjalmur Stefansson, *Unsolved Mysteries of the Arctic* (New York: Collier Books, 1962), 61.

3. M'Clintock, *The Voyage of the* Fox *in the Arctic Seas*, 119.

4. M'Clintock, *The Voyage of the* Fox *in the Arctic Seas*, 256.

5. M'Clintock, *The Voyage of the* Fox *in the Arctic Seas*, 258.

6. Steffanson, *Unsolved Mysteries of the Arctic*, 78.

7. James Morris, *Farewell the Trumpets* (New York: Harcourt, Brace 1973), 55.

CHAPTER 22

1. George S. Nares, *The Official Report of the Recent Arctic Expedition* (London: John Murray, 1876), 32.

2. Nares, *The Official Report of the Recent Arctic Expedition*, 35.

3. George S. Nares. *Narrative of Voyage to Polar Sea During 1875–6*, (London: Low, Marston, Searle, and Rivington, 1878), 2:93.

CHAPTER 23

1. Otto Sverdrup, *New Land* (London: Longmans, Green and Company, 1904), 1:1

2. Sverdrup, *New Land*, 2:267.

3. Derek Hayes, *Historical Atlas of the Arctic* (Vancouver: Douglas & McIntyre, Ltd., 2003), 154.

4. Robert E. Peary, *Nearest the Pole* (New York: Doubleday, Page and Company, 1907), 240.

CHAPTER 24

1. William R. Hunt, *Stef: A Biography of Vilhjalmur Stefansson* (Vancouver: University of British Columbia Press, 1986), 97.

2. Vilhjalmur Stefansson, *The Friendly Arctic: The Story of Five Years in Polar Regions*, (New York: The MacMillan Company, 1921), 307.

3. Stefansson, *The Friendly Arctic*, 236

4. Derek Hayes, *Historical Atlas of the Arctic* (Vancouver: Douglas & McIntyre, Ltd., 2003), 183.

5. Personal communication, Natural Resources Canada, National Air Photo Library.

CHAPTER 25

1. J. Bronowski and B. Mazlish, *The Western Intellectual Tradition*. (New York: Harper & Brothers, 1960), 55.

2. Vilhjalmur Stefansson, *The Friendly Arctic: The Story of Five Years in Polar Regions*, (New York: The MacMillan Company, 1921), 723.

3. Vilhjalmur Stefansson. *Unsolved Mysteries of the Arctic*. (New York: Collier Books, 1962), 115.

4. Stefansson, *Unsolved Mysteries*, 12.

5. Coolie Verner and Frances Woodward, *Explorer's Maps of the Canadian Arctic, 1818–1860* (Toronto: B. V. Gutsell, Dept. of Geography, York University, 1972), 75, Monograph no. 6.

6. Trevor H. Levere, *Science and the Canadian Arctic* (Cambridge: Cambridge University Press, 1993), 2.

BIBLIOGRAPHY

Agnew, T., A. Lambe, and D. Long. "Estimating Sea Ice Area Flux Across the Canadian Arctic Archipelago Using Enhanced AMSR-E." *Journal of Geophysical Research.* 113 (C10011) (2008), doi: 10.1029/2007JC004582.

Alley, Richard B. *The Two-Mile Time Machine: Ice Cores, Abrupt Climate Change, and Our Future.* Princeton: Princeton University Press, 2000.

Armstrong, Alexander. "Observation on Naval Hygiene and Scurvy, More Particularly as the Latter Appeared During the Polar Voyage." *Quarterly Journal of Practical Medicine and Surgery* 22 (1858): 295–305.

Asher, George M. *Henry Hudson the Navigator: The Original Documents in Which His Career is Recorded, Collected, Partly Translated, and Annotated.* New York: Burt Franklin, printed for the Hakluyt Society, 1860.

Back, George. *Narrative of the Arctic Land Expedition to the Mouth of the Great Fish River and Along the Shores of the Arctic Ocean in the Years 1833, 1834, and 1835.* London: Adamant Media Corporation, Elibron Classics Replica Edition, 2005.

Barrow, John. *A Chronological History of Voyages into the Arctic Regions.* London: John Murray, 1818.

Barrow, John. *Voyages of Discovery and Research Within the Arctic Regions From the Year 1818 to the Present Time,* London: John Murray, 1846.

Bartholomew, Michael. "James Lind and Scurvy: A Revaluation." *Journal for Maritime Research, British National Maritime Museum* (January 2002).

Berton, Pierre. *The Arctic Grail: The Quest for the Northwest Passage and the North Pole, 1818–1909.* New York: Viking Penguin, 1988.

241

Best, George. *The Three Voyages of Martin Frobisher: In Search of a Passage to Cathaia and India by the North-West, A.D. 1576–8.* Reprinted from 1867 ed. of *Hakluyt's Voyages* with Introduction by Sir Richard Collinson. New York: Burt Franklin, 1963.

Betts, Jonathan, *Time Restored: The Harrison Timekeepers and R. T. Gould, the Man Who Knew (Almost) Everything.* Oxford: Oxford University Press, 2006.

Bolles, Edmund B. *The Ice Finders: How a Poet, a Professor, and a Politician Discovered the Ice Age.* Washington, DC: Counterpoint, 1999.

Brown, Lloyd A. *The Story of Maps.* New York: Dover Publications, Inc., 1977.

Bronowski J., and B. Mazlish, *The Western Intellectual Tradition.* New York: Harper & Brothers, 1960.

Burnett, D. Graham. *Masters of All They Surveyed: Exploration, Geography, and a British El Dorado.* Chicago: University of Chicago Press, 2000.

Cooke, Alan, and Clive Holland. *The Exploration of Northern Canada: 500 to 1920: A Chronology.* Toronto: The Arctic History Press, 1978.

Cumming, W. P., R. A. Skelton, and D. B. Quinn, *The Discovery of North America.* New York: American Heritage Press, 1972.

David, Andrew C. F. "James Cook." In *Oxford Dictionary of National Biography.* New York: Oxford University Press, 2004.

Davis, John, Albert Hastings Markham, John Jane, Charles Henry Coote, and Edward Wright. *The Voyages and Works of John Davis, the Navigator.* London: Hakluyt Society, no. 59, 1880.

Dear, I. C. B., and Peter Kemp, eds. *The Oxford Companion to Ships and the Sea.* New York: Oxford University Press, 2006.

Delgado, James P. *Across the Top of the World: The Quest for the Northwest Passage.* New York: Checkmark Books, 1999.

Earle, Peter. *Sailors: English Merchant Seaman 1650–1775.* London: Methuen, 1998.

Energy, Mines and Resources Canada. *Canada Exploration 1497–1650.* The National Atlas of Canada, 5th ed. Ottawa, 1991.

Fardy, B. D. *John Cabot: The Discovery of Newfoundland.* St. John's Newfoundland: Creative Publishers, 1994.

Firstbrook, Peter. *The Voyage of the Matthew: John Cabot and the Discovery of North America.* London: BBC Books, 1997.

Fleming, Fergus. *Barrow's Boys.* New York: Atlantic Monthly Press, 1998.

Franklin, John. *Narrative of a Second Expedition to the Shores of the Polar Sea in the Years 1825, 1826, and 1827* (includes John Richardson, *Account of the Progress of a Detachment to the Eastward*). Rutland: Charles E. Tuttle. 1971. First printed 1828.

Franklin, John. *Narrative to the Shores of the Polar Sea, 1819, 1820, 1821, and 1822.* London: John Murray, 1828.

Gay, Peter. *The Enlightenment: The Rise of Modern Paganism* New York: W. W. Norton, 1966.

Goetzmann, W. H., and G. Williams. *The Atlas of North American Exploration.* New York: Prentice Hall, 1992.

Gound, Rupert T. *The Marine Chronometer, Its History and Development.* London: J. D. Potter, 1923.

Green, Meg. *Jacques Cartier: Navigating the St. Lawrence River.* New York: Rosen Publishing Group, 2004.

Hakluyt, Richard. *Voyages in Search of the North-West Passage.* London: Cassell and Company, 1886.

Hakluyt, Richard, and Edmund Goldsmid. *The Principal Navigations, Voyages, Traffiques, and Discoveries of the English Nation*. London: E. & G. Goldsmid, 1885.

Harley, J. B. "Deconstructing the Map," *Cartographica* 26 (1989): 1–20.

Harrisse, Henry. *The Discovery of North America: A Critical, Documentary, and Historic Investigation*, Amsterdam: N. Israel, 1961. First published 1892.

Hayes, Derek, *Historical Atlas of the Arctic*. Vancouver: Douglas & McIntyre, Ltd., 2003.

Hayes, Derek. *America Discovered: A Historical Atlas of North American Exploration*. Vancouver: Douglas & McIntyre, Ltd., 2004.

Hearne, Samuel. *A Journey From Prince of Wales's Fort in Hudson's Bay to the Northern Ocean*. New York: Da Capo Press, 1968.

Hood, Robert. *To the Arctic by Canoe 1819–1821: The Journal and Paintings of Robert Hood, Midshipman with Franklin*. Montreal: McGill-Queens University Press, 1974.

Hunt, William R. *Stef: A Biography of Vilhjalmur Stefansson*. Vancouver: University of British Columbia Press, 1986.

Gilbert, George, and Christine Holmes, eds. *Captain Cook's Final Voyage: The Journal of Midshipman George Gilbert*. Honolulu: University Press of Hawaii, 1982.

Johnson, Donald S. *Charting the Sea of Darkness: The Four Voyages of Henry Hudson*. Camden, Maine: International Marine, 1993.

Jones, J. W., *Divers Voyages Touching the Discovery of America and the Islands Adjacent*. First series, no. 7 (reprinted). New York: Burt Franklin, 1850.

Kane, Elisha Kent. *Arctic Explorations: The Second Grinnell Expedition in Search of Sir John Franklin*. Philadelphia: Childs and Peterson, 1856.

Leacock, Stephen. *The Mariner of St. Malo: A Chronicle of the Voyages of Jacques Cartier*. Toronto: Glasgow, Brook & Co., 1920.

Levere, Trevor H. *Science and the Canadian Arctic*. Cambridge: Cambridge University Press, 1993.

Markham, Albert H. *The Voyages and Works of John Davis the Navigator*. New York: Burt Franklin. Reprinted, originally published by the Hakluyt Society, 1970.

Markham, Clements R., and William Baffin. *The Voyages of William Baffin, 1612–1622*. London: The Hakluyt Society, 1881.

May, William E. *A History of Marine Navigation*, Henley-on Thames: G. T. Foulis and Co., 1973.

M'Clintock, Francis L. *The Voyage of the* Fox *in the Arctic Seas: A Narrative of the Discovery of the Fate of Sir John Franklin and His Companions*. Boston: Ticknor and Fields, 1860.

McGoogan, Ken. *Fatal Passage: The True Story of John Rae, the Arctic Hero Time Forgot*. New York: Carroll and Graf Publishers, 2002.

McGoogan, Ken. *Lady Franklin's Revenge*. Toronto: Harper Collins Publishers, Ltd., 2006.

Miller, Theodore R. *Graphic History of the Americas*, New York: John Wiley & Sons, Inc., 1969.

Morison. Samuel E. *The European Discovery of America: The Northern Voyages A.D. 500–1600*. New York: Oxford University Press, 1971.

Morison, Samuel E. *The Great Explorers: The European Discovery of America*. New York: Oxford University Press, 1978.

Morris, James. *Heaven's Command: An Imperial Progress*. New York: Harcourt, Brace, Jovanovich, 1973.

Morris, James. *Farewell the Trumpets: An Imperial Retreat*. London: Faber and Faber, 1978.

Mowat, Farley. *Ordeal by Ice*. Toronto: McClelland & Stewart, Ltd., 1960.

Nares, George S. *The Official Report of the Recent Arctic Expedition*. London: John Murray, 1876.

Nares, George S. *Narrative of Voyage to Polar Sea During 1875–6*. London: Low, Marston, Searle, and Rivington, 1878.

Neatby, Leslie H. *The Search For Franklin*. New York: Walker & Company, 1970.

Newby, Eric. *The Last Grain Race*. Hawthorn, Australia: Lonely Planet Publications, 1999.

Nickerson, Sheila. *Disappearance: A Map*. New York: Doubleday, 1996.

Nickerson, Sheila. *Midnight to the North: The Untold Story of the Woman Who Saved the Polaris Expedition*. New York: Jeremy P. Tarcher/Putnam, 2002.

Osborn, Sherard. *Stray Leaves From an Arctic Journal: Eighteen Months in the Polar Regions in Search of John Franklin's Expedition, in the Years 1850–51*. New York: George Putnam, 1852.

Osborn, Sherard, *The Discovery of the North-West Passage by H.M.S. Investigator, Captain R. M'Clure, 1850, 1851, 1852, 1852, 1854*. London: Longman, Brown, Green, Longmans, and Roberts, 1857.

Parry, William Edward. *Journal of a Second Voyage for the Discovery of a North-West Passage*. London: John Murray, 1824.

Parry, William Edward. *Journal of a Third Voyage for the Discovery of a North-West Passage*. London: John Murray, 1826. Reprinted from London: Feather Trail Press, 2010.

Parry, Edward. *Memoirs of Rear Admiral Sir W. Edward Parry*. London: Longman, 1858.

Payne, E. J., and C. R. Beazley. *Voyages of the Elizabethan Seamen*. London: Oxford University Press, 1907.

Peary, Robert E. *Nearest the Pole: A Narrative of the Polar Expedition of the Peary Arctic Club in the S.S. Roosevelt, 1905–1906*. New York: Doubleday, Page and Company, 1907.

Portinaro, Pierluigi, and Franco Knirsch. *The Cartography of North America, 1500–1800*. Edison, NJ: Chartwell Books, 1987.

Quinn, David B. "John Cabot." In *Oxford Dictionary of National Biography*. New York: Oxford University Press, 2004.

Richardson, John. *Arctic Searching Expedition*. New York: Harper and Brothers, Publishers, 1852.

Richardson, John. *The Last of the Arctic Voyages*. London: Lovell Reeve, 1855.

Ross, John. *Narrative of a Second Voyage in Search of a Northwest Passage, and of a Residence in the Arctic Regions During the Years 1829, 1830, 1831, 1832, 1833*. London: A. W. Webster, 1835.

Rundall, Thomas. *Narratives of Voyages Toward the North-West*. London: Hakluyt Society, 1849.

Sale, Richard. *Polar Reaches: The History of Arctic and Antarctic Exploration*. Seattle: Mountaineers Books, 2002.

Savours, Ann. *The Search for the North West Passage*. New York: St. Martin's Press, 1999.

Schwass-Bueckert, Kate. "Team Finds Lost 1800's Ship in Arctic." QMI Agency, appeared in *Toronto Sun* July 28, 2010.

Simpson, Thomas. *Narrative of the Discoveries on the North Coast of America, Effected by the Officers of the Hudson's Bay Company During the Years 1836–39*. London: Richard Bentley, 1843.

Simpson, Alexander. 1845. *The Life and Travels of Thomas Simpson, the Arctic Discoverer*. London: Richard Bentley, 1845.

Sobel, Dava. *Longitude: The True Story of a Lone Genius Who Solved the Greatest Scientific Problem of His Time*. New York: Walker, 1995.

Stefansson, Vilhjalmur. *The Friendly Arctic: The Story of Five Years in Polar Regions*, New York: MacMillan Company, 1921.

Stefansson, Vilhjalmur. *Unsolved Mysteries of the Arctic*. New York: Collier Books, 1962.

Sutherland, Peter C. *Journal of a Voyage in Baffin's Bay and Barrow Straits in the Years 1850–1851, Under the Command of Mr. William Penny*. London: Longman, Brown, Green and Longmans, 1852.

Sverdrup, Otto. *New Land*. London: Longmans, Green and Company, 1904.

Taylor, E. G. R. *The Haven Finding Art: A History of Navigation from Odysseus to Captain Cook*. New York: Abelard-Schuman, Ltd., 1957.

Taylor, Iain C. "Official Geography and the Creation of Canada." *Cartographica* 31 (1994): 1–15.

Thomson, George M. *The North-West Passage*. London: Secker & Warburg Ltd., 1975.

Vail, Philip. *The Magnificent Adventures of Henry Hudson*. New York: Dodd, Mead and Company, 1965.

Verner, Coolie, and Frances Woodward. *Explorer's Maps of the Canadian Arctic, 1818–1860*. Toronto: B. V. Gutsell, Dept. of Geography, York University, 1972.

Waters, David. *The Art of Navigation in Elizabethan and Early Stuart Times*. Greenwich, England: National Maritime Museum, 1978.

Watson, Harold Francis. *The Sailor in English Fiction and Drama 1550–1800*. New York: Columbia University Press, 1931.

Wilford, John Noble. *The Mapmakers*. New York: Alfred A Knopf, 1981.

Williams, Glyn. *Voyages of Delusion: The North-West Passage in the Age of Reason*. London: Harper Collins, Publishers, 2002.

Wood, K., and J. E. Overland. "Accounts from 19th Century Canadian Arctic Explorers, Logs Reflect Present Climate Conditions." *Eos* 84 (2003): 410–12.

Wroth, Lawrence C. *The Voyages of Giovanni da Verrazzano, 1524–1528*. New Haven: Yale University Press, 1970.

INDEX